Space Anomalies

By Steve Preston

1st Edition

© Copyright 2017, Steve Preston
All rights reserved.
No part of this book may be reproduced, stored in a retrieval system, or transmitted by any means, electronic, mechanical, photocopying, recording, or otherwise, without written permission from the author.

Table of Contents

SPACE ANOMALIES 1
TABLE OF CONTENTS 3
INTRODUCTION 5
FORMING OF THE PACIFIC OCEAN ANOMALY 11
MARS ENCOUNTER ANOMALY 17
DEAD PLANETS ANOMALY 23
EXTINCTION ANOMALY 28
NO MEMORY ANOMALY 30
ARE FLYING SAUCERS REAL ANOMALY 34
LUNAR COLONIZATION 37
LUNAR BUILDING ANOMALY 38
LUNAR TOWER ANOMALY 46
TOWERS IN CRATERS ANOMALY 49
MACHINERY ANOMALIES 51
LUNAR ROADS ANOMALY 55
LUNAR BLAST ANOMALY 59
TINY NOSE ANOMALY 60
HOLLOW MOON ANOMALY 65
PEOPLE UNDERGROUND ANOMALY 68
VENUS BIBLE ANOMALY 72
VENUS COLONIZATION AND DESTRUCTION 76
VENUS VISIT ANOMALY 77
HOW BAD IS VENUS? 83
VENUS CRACK ANOMALY 87
BLAZING TAIL ANOMALY 89
CAROLINA BAYS ANOMALY 97
TRAVEL ANOMALY 100
BIBLE DESCRIBES THE PLEISTOCENE WAR 101

BURST OF RADIATION ANOMALY	103
LAVA AND MOUNTAINS	111
VENUS RIVER ANOMALY	113
BUILDINGS	117
ROADS ANOMALY	120
VENUS WAR ANOMALIES	122
BOMB BLASTS	124
VENUS LIFE ANOMALY	131
MARS COLONIZATION ANOMALY	134
OLD MARS ANOMALY	135
MARTIAN WATER ANOMALY	140
INDUSTRIAL ANOMALIES	146
ANCIENT CITIES ANOMALY	151
MARTIAN BUILDINGS ANOMALY	154
TOWER ANOMALIES	158
MARTIAN ROAD ANOMALIES	160
ANOMALOUS ARTIFACTS	164
MARTIAN FLIGHT ANOMALY	168
MARTIAN LIFE ANOMALY	173
MARTIAN TREES ANOMALY	176
MARTIAN WAR ANOMALY	180
TUNNEL ANOMALIES	184
ARE PEOPLE THERE NOW?	189
ANOMALIES BEYOND MARS	194
CONCLUSIONS AND DISCUSSIONS	201
ABOUT THE AUTHOR	205

Introduction

The Earth was not seeded by a foreign group of humans from a starship nor did the earth populations start by having piles of sugar on stumps bond precisely in a string of DNA at the same time hundreds of other piles did the same thing so there would be a viable colonies to initiate life. That being said along the "development stages" of human life there has been external control by our Creator God and by subsequent creations of that God on Earth.

Some tell you, life on Earth is a chain of evolutionary mutations and sequential development from Cavemen to the atomic age. By forcing this insanity, thousands and thousands of details and artifacts had to be place on the sacrificial alter of anomaly never to be used to understand the truth. This book will describe some of these forbidden facts and how twisted the details of these facts have become.

This is not a physics book concerning the development and existence of our universe. Instead, it uncovers and explains anomalies associated with our closest relatives in this solar system; Lunar, Venus, and Mars as **these neighbors hold evidence needed for us to understand the truth**. The problem is that the insanity of coverup and the complex reasonings needed to assure support for a favored theory and comfortable definition has gone on so long, no one will fight for our children to allow them to get a truer understanding of our nearest neighbors and how we have been associated with them in the past. It is criminal how quasi-scientific textbooks describe things as Physical Laws or Truths, when many of their observations are no more than a conjecture that should have been thrown away long ago. It is so bad, even the sightings of these UFOs are thrown into the "Twilight Zone" area and when evidence of life on our neighboring planets is seen, the objects are quickly debunked by some crazy notion. After things are stated over and over, it is difficult to turn even the worst lie around. To make things worse, those investigating alternative theories that don't require anomalies to satisfy a premise are called "FRINGE Scientists" with a sneer as they say the word "fringe". Hopefully, this book will begin your journey and actually allow the opening your eyes to other probabilities. To help, I have collected a tiny fraction of the details of our nearest neighbors. If you have just been allowing yourself to be dragged around by conformists so afraid of rocking the boat that they tell you lie after lie after half-truth after half-truth and bundle it into what seems to be reasonable. That is until you really think about what they say and how many thousands of pieces of evidence they THROW AWAY so people don't get confused.

What You Think You Know

You are not the blame for closeminded inappropriate concepts that have been drummed into you heads over and over.

- You think our near planets had no part in the development of our Ecosystem. [Lie!]
- You think you know people didn't live on the near planets. It sounds absurd and too science fictiony. [Lie!]

- You either think religion and reality cannot share the same space or you don't think the Biblical Stories describe our planetary neighbors except in a general way. [Lie!]
- You may think you have a pretty good handle on how the earth is the way it is and what the planets are like, but after reading this book, hopefully, you will see that you have not been told the truth or at the very least, information you needed to make a reasonable attempt at the truth has not been provided to you---until now.
- Some of you think planetary seeding from an ancient race is the answer, but the delicate nature of the timing of our universe won't allow for civilization back far enough to support the notion.
- Hundreds of modern human footprints alongside dinosaurs, finding 16 super ancient nuclear reactors with missing fuel, evidence of electrical devices in geodes and stone, petrified giant skull-bones, evidence of ancient flying, and complex manufacturing are all simply dismissed or pushed to the "visitors from Alpha Centauri seeding block. These people lived with dinosaurs and were not from another world and sometime in our very ancient past, these people established colonies on our nearest neighbors.

When did humans become civilized? *If you believe they became civilized with the beginning of the Bronze Age, that's not far enough back. If Neanderthals is your answer, you are still way off. If you believe that Adam was the first civilized human, again you are wrong and you have not read the Bible Genesis story completely. While the evidence is overwhelming and convincing, some will have a hard time with this major factor in earth's development.*

Where do UFOs come from? *If you think some time travel or travel from a distant galaxy, there is an extremely high probability that you are way off. UFOs are driven by humans from Earth.*

When did people learn to fly? *If you are thinking Wright Brothers or even the flying balloons of the Roman and later times. Sadly, you are way off as much of our historical details have been removed from textbooks.*

Why do UFO pilots look human and have small bodies, small noses, big eye, and big heads? *The answer has nothing to do with being from another planet? If your answer is that UFO sightings are simply mass hysteria or told by lying fools, you again have not received the science and history you need to understand this world.*

Basic Lunar Questions

Did people live on the moon at one time? *I'm not talking about little green men, I'm talking about normal people.*

Where did the moon come from and why does it have such a low density? *Some may know that when the Pacific Ocean was made, one chunk became the moon, but what really happened?*

Are there ancient texts *that help us understand something about the moon? If they exist, can they be cross compared to look for embellishment, misunderstanding, or lies similar to what has been inserted in our "quasi-science textbooks?*

Can the moon sustain life? *First, the moon has almost no atmosphere and no air, so on might think this is impossible, but there are ways to counter this and NASA has been working on a number of them. Also, it should be known that*

Lunar is filled with massive caverns that could hold air, water, and comfort.

Basic Mars Questions

How did Mars get split in half? *You might be saying, "If Mars had been torn in half, you would have heard about it before now." Unfortunately, the answer is you SHOULD HAVE been told about it.*

Did people live on Mars at one time? *It is comfortable to say no, but the facts don't go along with the comfortable answer and the massive amount of trapped water seeping to the surface tell us water and air to support life can be provided. Also, miles of underground tunnels that have been found at areas where they breech the surface suggest someone might have used the tunnels at one time.*

Why are radioactive fallout *materials found on Mars?*

Is water on Mars? *What was it like in ancient times?*

Could Mars have supported human life *during ancient times?*

What does Mars have to do with mountains *on our Planet and the Pacific Ocean?*

Are there roads on Mars? *If so, what were they for?*

Basic Venus Questions

How and when did Venus turn into a ball of fire? *If you believe that a thing we call the Greenhouse Effect affected Venus---You have been fooled by those with inappropriate agendas. If you believe that Venus caught on fire millions of years ago you are wrong again.*

Why is methane on Venus? *Doesn't it come from animals?*

Are there ancient texts that help us understand something about Venus/ If they exist, can they be cross compared to look for embellishment, misunderstanding, or lies similar to what has been inserted in our "quasi-science textbooks.

When did the Huge split that almost split Venus in half occur- *You probably are thinking, "There is no such split!" While no one has told you about it, the huge split is important to earth's history.*

What are the Carolina Bays *and what do they have to do with Venus and Venus Flytraps?*

Did people live on Venus at one time? *It is comfortable to say no, but the facts don't go along with the comfortable answer.*

Is there life on Venus now? *With 700- degree temperatures and unbearable air pressures, the easy answer us no, but we should possibly look at the evidence.*

Why did people see Venus with a blazing tail *during ancient times and write about it?*

Are there roads on Venus? *If so, what were they for?*

Are there melted buildings on Venus? *If so were they occupied at one time?*

What is strange about Venus Flytraps?

These and many other anomalous details about our near neighbors is covered in the book. Let's first look at how the Pacific Ocean was made and how Prestonia Disappeared.

Forming of the Pacific Ocean Anomaly

If you're looking for a Science book that ignores the more unpleasant things that have been witnessed either as physical evidence or by first hand descriptions and knowledge, you had better just shut the book right now. If you are looking for a book that will comfort you, you also have come to the wrong place. If you really want to learn about how the earth got to be the way it is today regardless of what you were told in the past, please continue. You will find that the lies and half-truths that you have been told don't change the past, they cover it up.

While this section is not directly connected to planetary anomalies, we need back out of some of the more egregious quasi-science, and comfort-histories statements and provide some level of explanation about why they are rubbish. The things I'm talking about are as follows:

Bode's Law of Planetary stability- total rubbish as there are many signs that even 11 thousand years ago major shifts in the planetary locations were still going on.

Tectonic creation of the massive mountain ranges from Antarctica, through South America, Central and North

America, and down the other side of the Pacific Ocean is crazy. The series of Mountains go well over ½ way around the Pacific Ocean, all plates would have to be pushed away from some massive volcano thing in the center of the Ocean that just disappeared as shown below.

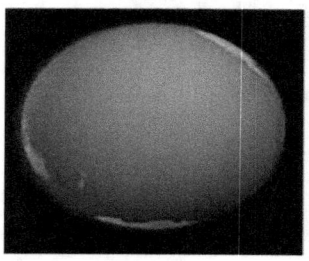 

The idea that there was only one super continent [Pangea] all clumped up on one side of the Earth and the Pacific Ocean hole with almost no crustal material at its bottom was not part of a massive continent that was sucked into space from some ancient catastrophe seems pretty much dumb. The following "Earth curstal thickness image" shows the massive crustal thinning from the loss of "the missing continent, Prestonia" at the end of the Permian Age.

The notion that people lived like Stone Age people during the Pleistocene and the remanufactured, 20-thousand-year-old radio-active, unfossilized, T-Rex and other dinosaurs and signs of flying machines are all anomalies.

The notion that no humans lived during the Cretaceous and hundreds of footprints, shoeprints, indications of humans walking on the same beaches as massive dinosaurs, complex manufactured goods being found in geodes, and deep in coal mines, and the 16 nuclear reactors in the Oklo mountains that were operational during the Cretaceous--- all are just anomalies. Samples of fossilized shoe and foot images shown next.

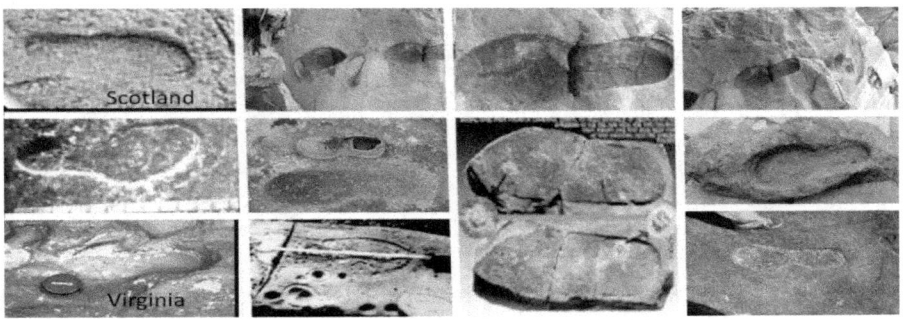

The reliance on nuclear decay as a reasonable timing source is irresponsible. Scientists know that even a small neutrino-loaded sunspot or a super-heated volcano, or any nuclear reaction can change the "apparent timing by a 1000% and the erroneous timing has been substituted by the cross-comparison of Ice core direct CO_2 cyclic measurement, O_{18} concentration cycles, archeo-magnetic cycle sampling from the mid-Atlantic basin lava solidifying magnetic markers, and the earth spin changing hot-spot tracking of the Hawaiian Island chain as its location describes earth's historical equatorial positioning. The first graph in the following collage shows the CO2 cyclic nature for Antarctic Ice coring on top followed by the magnetic field shifting cyclic nature, followed by the, O_{18} density cycling [major component in seashells], followed by the track of the Hawaiian Island Hot spot knowing the equator is perpendicular to the hot spot motion. Everything lines up to allow us to determine a less arbitrary date supplied by Nuclear decay. Put together we get a MUCH better understanding of when things really happened, especially major atmospheric shifts during the various extinction periods. From it we get a much more correct Standard Geological Timeline shown to the right.

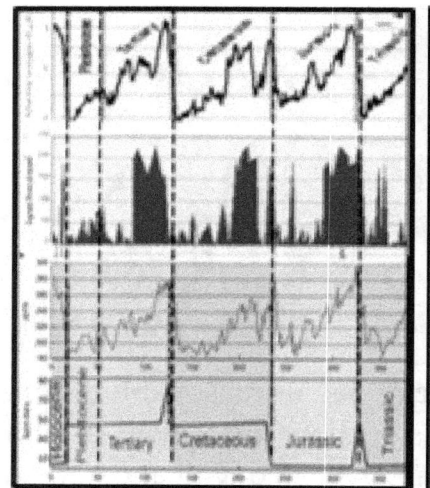

Standard Geological Timeline		
Era/Period/Epoch	Time (M yrs. ago)	Time (T yrs. ago)
Archaeozoic Period	5000-1500	50,000-3000
Proterozoic Period	1500-545	3000-1000
Cambrian period	550-500	1000-900
Ordovician period	500-440	900-800
Silurian period	440-410	800-700
Devonian period	410-365	700-600
Carboniferous	365-300	600-500
Permian period	300-250	500-400
Triassic period	250-212	400-300
Jurassic period	212-145	300-200
Cretaceous period	145-65	200-100
Tertiary period	65-1.8	100-40
Pleistocene period	1.8-0.01	40-10
Holocene period	0.01-0	10-00

What makes anyone think the Mars and Earth catastrophe occurred at the end of the Permian? While there are a number of things that tells us about the near collision timing like the worst extinction period of all ages, but I want to show you something very interesting. The following is an extension of the Ice core sampling data from Antarctica as described above.

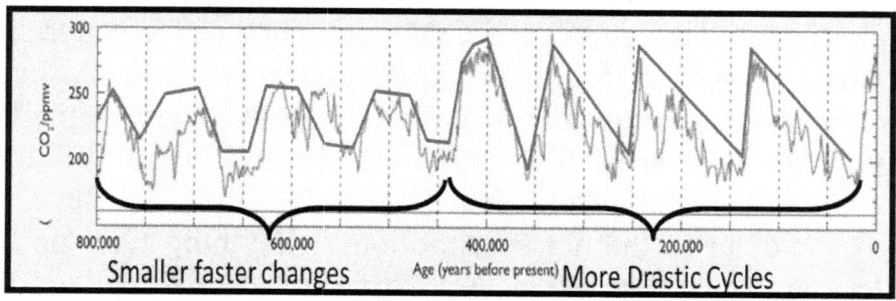

What you need to see here is the marked <u>difference in the characteristics of the Earth about 400 thousand years ago</u>. The larger changes in atmospherics after that time and much less violent cycles before that time. Just about anyone can tell some MAJOR change occurred as our planet became lopsided whenever the supercontinent "Prestonia" was sucked into space had had a long-lasting increase in Atmospheric shifting. Don't let anyone tell you the massive change in CO2 variations was caused by people driving automobiles or using coal to heat their homes. Global atmospheric changes are not dictated by ANYTHING humans do. We could not stop the cyclic nature of climate change if we wanted to.

The notion that Venus began to boil from greenhouse effect millions of years ago all on its own is simply a way to scare the masses. Everything about that is wrong. We now know the entire surface of Venus appears NEW. The riverbeds are still in place, the lava flows have simply halted abruptly and are still well defined, and there are the Carolina Bays and Venus Flytraps which we will discuss later, but whatever happened on Venus did not happen long ago. We even know the catastrophe happened about 11 thousand years ago and massive amounts of details were recorded of its demise.

The idea that we can explain away anything we want to by saying little green men did it as no one really understands what that is about anyway. Unfortunately, logistics of time space travel over and over to the same spot to visit the same single planet in the universe is way too difficult to accomplish. We are flying through space at something like 2,000 miles per second making all types of gyrations around galaxies, our sun and between galaxy clusters. Even if someone could get back to nearly the same TIME, finding the space is another matter and there is the annoying requirement for a cognizant observer to view an event or it disappears according to the modern quantum fluctuating matter sciences. Besides UFO traffic has been witnessed for thousands of years so it would be a full-time, lifetime job for a massive crew to attempt this "sightseeing" thousands and thousands of times with no real reason unless we were the only food source and their planet had burned up. -------- If you don't find out anything else, please consider humans from Earth as the MYSTERIOUS aliens. ---- Today, these humans must live either on the earth or nearby to visit us by the thousands as we have noticed for the past few thousand years.

The big question right now might be, what happened to our Earth 400 thousand years ago and will it happen again. Whatever it was, the Earth shifted on its axis, almost all life died, the super continent of Prestonia disappeared and in its place a massive hole with almost no crustal lining its base, even today.

Mars Encounter Anomaly

You are taught that plate tectonics made the mountain range that goes around the Pacific, but if that were so; how come all the Earth crust is missing where the Pacific Ocean is today? If you are thinking it was a lie, there is hope for you.

Here is the anomaly; 1/3 of the Earth and ½ of Mars is missing. It is becoming more and more accepted that this massive loss of planetary material happened at the same time around the end of the Permian Period. By nuclear decay timing, this was about 250 Million years ago [+0-250 million]. By the new timing standards, this was about 400 thousand years ago.

The orbit Mars was more elliptical than its present one which allowed Mars to come close to Earth from time to time. It was fairly well known, Mars came so close to Earth that the Pacific Ocean was scooped out and half of Mars was yanked out at the same time from their mutual gravitational pulls, as shown in the following graphic.

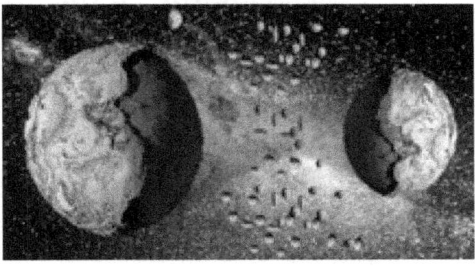 

The debris field shot out near the farthest portion of the previous Martian orbit to become the Asteroids, but one piece was captured by Earth and two smaller pieces were captured by Mars to make the moons of both planets. Also; the Martian orbit was brought in as it established its circular orbit. Soon the planets began healing themselves by pushing crustal material into the hole. On Earth, the Atlantic Ocean came about as Pangea, the only remaining super continent, split in half along the east coastline of the Americas. The preceding graphic, right shows the Earth as it is today, minus the water.

Mars is doing something similar, but now, almost all the surface water has disappeared today so it just looks like one hemisphere has massive numbers of meteor craters all over the place while the "ripped away" hemisphere only has those that have fallen over the past 400 thousand years. The incident left Both planets slightly closer to the sun making both planets, slightly warmer. Also, both planets began to spin faster as their effective diameters had reduced. From the faster spin, everything on Mars and Earth were lighter so animals grew larger, but the planets were spinning too fast to hold onto their atmospheres. Two additional events occurred on the Earth [Cretaceous Extinction and Pleistocene Extinction] which slowed the planet down before too much atmosphere was lost. Mars had no such control so the thriving planet, began losing its atmosphere and then its surface water. This took many thousands of years before finally stabilizing just a few thousand years ago.

I know this is not what you were told or what you read in you Earth Science books, but think about it.

- Tectonics could not have yanked out Prestonia and there is essentially no crustal material under the water.

- There is a reason why atmosphere samples of pre-Cretaceous Earth show much more Oxygen. -There used to be more atmosphere-
- There is a reason why the animals got so huge during the Mesozoic times, but then reduced back to "normal sizes". --Animals were lighter after the Earth spin increased—
- There was a reason massive numbers of animals became extinct at the end of the Permian. –The atmospheric and geographic trauma as the super continent was sucked away was simply too much for almost anything. --
- There is a reason why massive flying dinosaurs could fly and Diplodocus could raise his head high in the air and still be able to pump blood to the brain during the Mesozoic times. --They were lighter-
- There is a reason why the moon is made up of the same materials as Earth and its rotation is synchronized with Earth—It came from Earth--
- There is a reason Half of Mars has almost no crustal material similar to the Earth. – It almost hit the Earth-
- There is a reason why half of the planet Mars has so few craters. --The incident happened only 400 thousand years ago--
- There is a reason there are still signs of life on Mars- Mars was livable until very recently, and people used to be able to leave the planet--
- There is a reason we see where massive oceans used to be on the surface of Mars, but are now lost. --As it began spinning faster, it began losing its atmosphere and water—
- There is a reason the Marianas Trench along the edge of the Pacific Ocean and Valles Marineris Trench along the separating edge between the low and high halves of Mars look like someone was ripping the planet apart. – Both

planets were ripping each other apart, 400 thousand years ago.

The following graphic shows anomalous features of Mars. The southern half is filled with craters and 8000 meters higher while the northern half is not and along the edge of the dividing line we see a secondary rip called the Valles Marineris.

On Earth, we find a similar secondary rip along the joining line of the 2/3 of the planet with substantial crust and the 1/3 having almost no crustal material at the bottom of the Pacific Ocean.

Hopefully, you are starting to sense that what you were told about our near neighbors; Lunar, Venus, and Mars, might not be completely correct. Our first stop is to examine the anomalies of the moon and see why they might not really be anomalies, they simply are items that are ignored so that geologic historians can keep printing the same trash without thinking about what they are telling our children. Before we look at the various anomalies on our nearby planetoids, let me just give you an overview of what we will be addressing and some of what we will be seeing in upcoming chapters. One of the things to learn from this book is the Solar System is not "that stable" and in the past, it was much less stable. Four hundred thousand years ago, we find the close encounter of Mars and Earth and the massive explosion that created the Pacific Ocean, but it was not the only significant and horrible earth explosion. This first one occurred at the end of the Permian Age, but another explosion would rock the Earth again at the end of the Cretaceous Age. Not to be left out, a third massive explosion initiated the Pleistocene Extinction and worldwide flood. This last event happened 10 to 12 thousand years ago. Different that the Mars encounter, this one involved Venus and it was initiated by an electrical discharge. Let me explain this with a couple of diagrams.

The Earth, Mars, and Venusian orbits have not been very stable. While our obits were elliptical, Venus, Earth and Mars all were located in what is known as the habitable zone. All had somewhat similar characteristics including oceans, rivers, rolling hills and meadows, plants, and animal life. Humans were first found on Earth and they would venture out to their closest neighbors after a time.

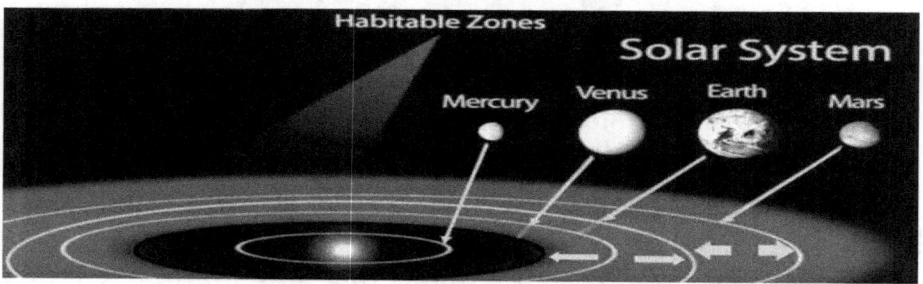

The instabilities and elliptical orbits have caused great destructions.

Great portions of Mars and Earth were yanked away and Venus almost split in half as the planets changed locations in our solar system to where they are today as shown next.

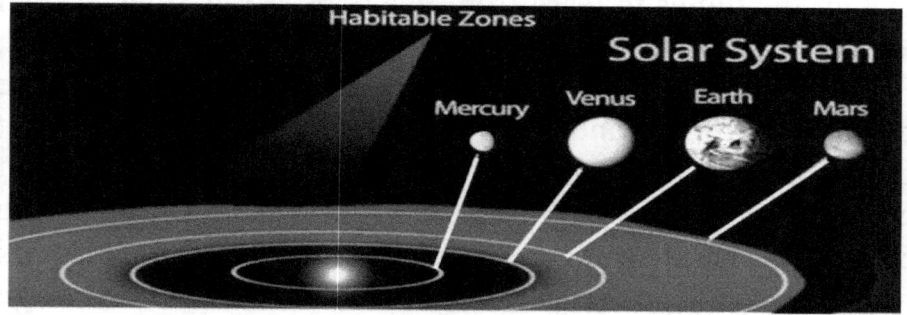

Today Earth is the most habitable planet in the Solar system, and our orbits are much more circular, but that does not mean all planets except Earth are dead.

Dead Planets Anomaly

We are told, our moon, Venus, and Mars have characteristics consistent with dead planetoids. More than that, they never have ever been colonized for any reason. Therefore, anything found on the Moon, Mars, and Venus [or any other planet] that appears to be manmade is an anomaly to be completely ignored.

There are good reasons for geologic historians taking this stand. The first, is no one wants to come out and say there are UFOs being seen. Secondly, if someone does want to think the thousands of UFO sightings are real, someone might be flying the things. Thirdly, if there were civilizations on these planets it completely destroys how they laid out their crazy ideas of uncontrolled Evolution, Stone Age stages of mankind, and the universe and all the millions of stars and planets simply being banged into being by some uncontrolled Cosmic Event.

Alien UFOs Don't Come From a Faraway Galaxy

Sorry about this section, but it makes you wonder if the writers of these textbooks even know about the new sciences that tell us reality is a very delicate thing where "cognizant observers" are required to turn what are called "Quantum Fluctuations" into matter and that if two observers are not linked together in some way, they will establish more than one reality. When that happens, people experiencing one reality will not be recognized or understood in the other reality.

Einstein's Dual Timing

Einstein is the one that started all the mess by essentially proving that if someone goes near the speed of light, time stops for him. He cannot age; when everyone else gets old around him as he is in a different "reality". Time passes and doesn't pass at the same time. Today, participatory quantum mechanics scientists go one step farther and require humans to be witnessing the universe or, they tell us, there is <u>no universe</u>. There is no universe and a universe at the same time.

This means the only reality is now.

Going back in time forces a different reality which is a great thing, because they have determined that the Big Bang event had to have occurred right on top of the Earth and there would be no solar system, no earth, and no sun at the epicenter of creation.

Hydrogen Turns to Helium

Another example is more to the point. When we look at stars near us they are mostly yellow, but distant stars still traveling at speeds over ½ the speed of light turn red. Dr. Hubble had a nice name for this "red shift", but this is what is happening. The stars are going so fast, their Hydrogen changes to Helium in our reality and at the same time it stays hydrogen. It is both at the same time. Again, I'm sorry for the ranting, but here is why I did it.

Aliens Are from Earth

As I said earlier, I am in no way suggesting that thousands of alien pilots are streaking across the universe to visit our planet over, and over, and, over, and over, and over again. Nor do I hold to the suggestion that humans were seeded by an alien race making it improbable that there is a God who designed humans in his triune image. I know there is an army of people now claiming such things, but it still doesn't make sense. Aliens in a different part of the universe would have a different reality and a different reference to time. Converting to ours might be possible, but certainly problematic and the hundreds of thousands of "visitations" would all cause shifts in reality. I'm not just talking about folding the universe as some suggest to travel great distances without light speed, I'm simply saying it's not just go from one planet to ours and back again and here and there and here and there and here and there over and over again while our earth is tumbling through space at about 20 thousand miles a second and the other planet is pushing through at 5 thousand miles a second with a completely different vector.

The other issue is one of the age of the universe in our reality. All the stars and everything in the universe we perceive is timed just right for carbon distribution needed for our development. A universe younger would not have enough carbon and an older one would have much larger molecules as the main building block for life. Aliens would not be more ancient than us in our reality. They would have to give something up.

Let me tell you a story that will be backed up later. Right now, it will sound a little crazy because you have not been introduced to much of the history, artifacts, science, religious testimony, comparative timing analysis, and other details that you should have learned in our less than perfect historical novels and quasi-science text books that gloss over anything that might make students uncomfortable or have to think.

When Were People on Earth?

As I stated, humans lived during the time of the dinosaurs. There simply is no reasonable way to deny that fact. Our Bible, many other ancient texts, and artifacts tell us that these first people [called *"giants of old"* in Genesis] were created by God and almost all died in a massive war that is marked by the Cretaceous Extinction. The Bible says that all the cities were destroyed and the Earth was as if it was without form and void. [in Genesis, Isaiah, and Jeremiah] That was a nasty war and all the dinosaurs died. There is a possibility that some colonization occurred back in that very ancient time, but it doesn't seem right to me and yes; the 16 unbelievably ancient nuclear plants found in Africa and known to have a substantial loss of fissionable material, were operational even during the Cretaceous time. We are talking about a highly civilized group of people inhabiting the entire world for over 100 thousand years and every year learning more about science, manufacturing, biology, and flying transport.

Who May Have Colonized Nearby Planetoids?

We are told some of these giant people survived into the Tertiary and finally into the Pleistocene and there is a very good reason to believe there were a substantial number of colonists who went to Venus, Mars and the Moon. Unfortunately, there was another massive war that was well under way 15 thousand years ago. We even know it was a nuclear war from 12-thousand-year-old nuclear byproducts, finding radioactive, unfossilized T-Rex bones, and a massive increase in mutations of human DNA just before the Pleistocene Extinction. It got worse and worse and we can believe at least a small amount of the fighting included the colonists or soldiers on the nearby planetoids. We are told specifically about battles on the planet known as Rahab [the Vain Planet]. Our Bible provides substantial details about this particular planet and its complete destruction.

War on Rahab

According to much physical evidence, about 11 thousand years ago, something happened to Venus. I know you were not told it happened so recently, but anyone examining the surface of Venus says the same thing; *the pristine character of the surface in such a hostile environment can only mean the destruction was a recent event.* Then there are craters. Along the Venusian Equator are more craters than those other parts, as if something nearby split apart and bits of debris hit as Venus was spinning. On Earth, we find over ½ million craters still extant on the East Coast of the United States. All were caused about 11 thousand years ago near the end of the Pleistocene. Venus flipped on its axis and slowed its rotation from the event and it almost instantly burned up everything. All the water evaporated, but there is still methane lingering and noble gases which further confirms Rahab died in the not too distant past. I know you are sitting there saying the word anomaly, but soon you we see the errors in children's textbooks.

Extinction Anomaly

We may never know what really caused the Earth Axis to shift 10 thousand years ago, but the massive bombardment of meteors possibly helped destabilize the Earth Atmosphere forcing the Pleistocene Extinction. From Archeo-magnetic, molten-metal alignments, hotspot tracking data, and simply looking at the line made by the invading meteors we can determine the Earth shifted about 30 degrees on its axis pushing over a million peaceful Mammoths, grazing in a meadow, into the Arctic circle, quickly freezing them with flowers still in their mouths. The previous Poles melted and were reestablished where they are today and the entire Earth flooded by action of enormous tidal-wave action. Even people living on mountain-sides were killed, but some escaped on massive boats, like Noah and in flying machines as described in ancient Sumerian texts. The outputs on Venus and Mars were burned up as we will see later so the only thing the people in the flying machines could do was to sit back in fear and cry out to God. On board Noah's ship, it was just as bad and they thought they would not make it as the whole world was shaking and they were tossed like "Pottage in a cauldron".

Sumerian *"Tablets of Enki"*- *An ark was built and the <u>seed of every animal</u> was transported along with the human survivors. <u>Above the Earth, the Annunaki [people in flying</u>*

machines] cried out in fear. Those saved from a worldwide flood sent out birds. God also sent a rainbow afterward. The Annunaki also survived.

Canonized Jewish Text-*"Jasher -6:11-32 And on that day, the Lord caused the whole Earth to shake, - and the whole Earth was moved violently, --and all the fountains in the Earth were broken up, --And the rain was still descending upon the earth, and it descended forty days and forty nights, and the waters prevailed greatly upon the earth; and all flesh that was on the ground or in the water died, whether men, animals, beasts, creeping things or birds of the air, and there only remained Noah and those that were with him in the ark. And the ark floated upon the face of the waters, and it was tossed upon the waters so that all the living creatures within were turned about like pottage in a cauldron.*

Harappa Flood-In ancient Pakistan, the Flood story was depicted as a boat full of animals [Shown next left]

Egyptian Flood-*Pyramid Texts* states *—The 3rd Period was called the Golden Age of man and a worldwide flood destroyed it.* In ancient Egypt, the Flood story was depicted as a boat full of animals as shown above right.

No Memory Anomaly

After everything settled down, civilizations again thrived. Ok! Venus was completely destroyed and Mars had lost almost all its oxygen by this time and the moon had never really been a great place to live, but we can believe there were still colonists on Mars and the Moon. On Earth things were "civilized and the biologist worked on DNA while the Engineers created new exciting machines. All of a sudden, things stopped 3100 BC. Humans, on Earth, became like animals and lost all memory of their past greatness.

Chinese Story-This comes from *"Shooikng"* -*The Maotse lost the capability to communicate but they previously talked to people from the sky. Later they could no longer go to the sky. They lost the memory of flying. When the Maotse brought trouble to the Earth, Changty [God] saw that his people had lost virtues and halted all communication between sky and Earth.*

Lower Congo Story-*The sun met the moon and threw mud, making it dimmer. -While the moon was dim, a huge flood occurred. -Men put their milk sticks behind them and were* <u>turned into monkeys</u>. *Later, a new race of men was created.*

Pre-Maya *"Popul Vuh"*-*The ancient ones had the power of understanding; They succeeded in knowing everything that* *could be seen or known in the world. They investigated the four corners of the heavens [experienced extended space travel] They investigated the round surface of the earth. Then one day the heart of heaven blew fog in their eyes, [Their brains lost abilities they once had.] They could not see clearly any more -This was the way that the wisdom and knowledge of these first people was destroyed.*

Essene Evidence-*Jasher 8:33-39- Those who said, we will ascend to heaven [people at the Tower of Babel during the Babel War] and serve our gods, became like apes.*

Totonac- Mexican Tradition-*After the flood, the boat finally rested and God reversed man's face and hind parts and turned him into a monkey.* [Probable indication of man losing memory of technology after the Babel incident]

Mayan Tradition-*During the second creation, people turned into monkeys and the world was destroyed by wind.* [Possibly talking about the destruction of Babel by wind and another indication of people having their brain functions erased]

Aztec History-*During the age of the four winds men turned into monkeys according to Codex "Laticano-Vatino"* [Possibly wind destroyed the tower and the monkey thing keeps coming back.]

Tibetan History- *"Tibet was almost totally inundated by the flood. The survivors had been little better than monkeys."*

From other texts, we fill in some parts. The Indians called it the Bharata War and others called it by different names, but the World again at war by about 3500 BC. Haplographic testing of DNA mutations shows that almost half of all human mutations occurred during this war. The once great societies of men were all in ruins and there was no longer any knowledge of flying machines, Space travel or colonization.

The modern human brain began to atrophy as shown next, modern human brains are smaller than either the Cro-Magnon or Neanderthal brains as a result of these mutations. Luckily, some we not on the Earth during the massive wars and nuclear fallout or biological agent that spread to all the people.

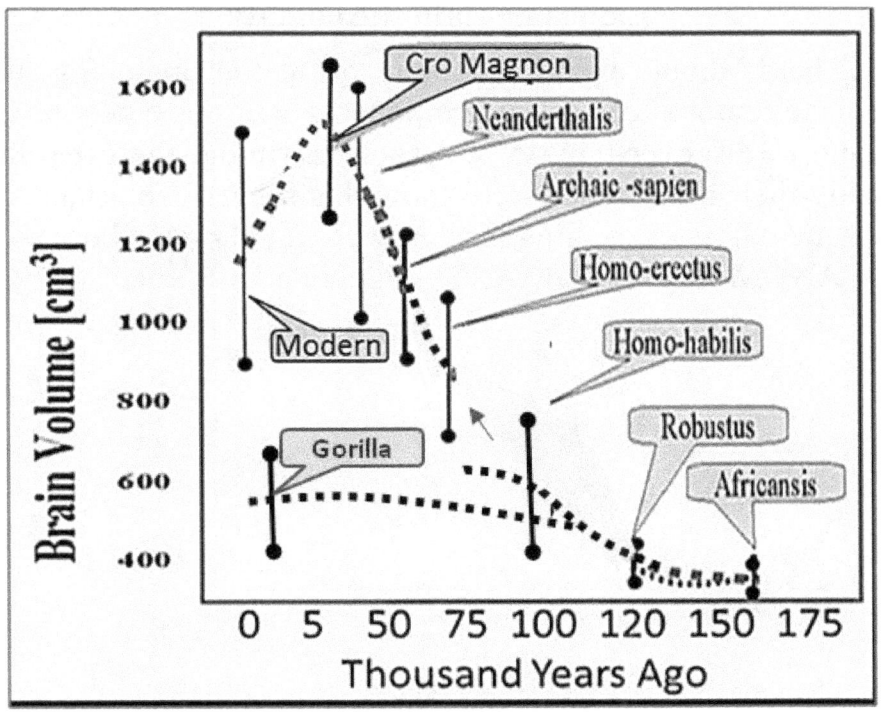

UFO Pilot Anomaly

While there is little proof the pilots of UFOs today are humans except for reason, there was a recorded interview a couple years ago with one of the pilots. In the interview, he continuously insisted he and his people came from Earth.

Strangely, while spiders on Earth look very different than we do, the pilot had the same number of eyes, noses, mouths, feet, arms, hands, heads, buttock, backbone, neck and all that one would conclude would not be alien, some still try to insist they are not human. In general, their eyes have grown a little larger and their bodies have reduced in bulk, but there simply are no non-human traits reasonably documented. It was as if they have been living in an area with less light and less gravity like underground on the moon or Mars.

Experimentation Anomaly

We here about experimentation on cattle including the delicate removal of genital components as if these people are trying to discover how to re-associate without the offspring losing their brain function. I know this seems like a fantasy, but guess how everything is addressed in your texts books-------- ANOMALY, ANOMALY--- ignore all evidence.

Are Flying Saucers Real Anomaly

For those thinking all this flying saucer stuff is nutty, I suppose you know about various versions made by the Germans in the mid to late 1940s as shown below. Form these initial attempts, Russia, America, France, and Canada all developed Flying saucers.

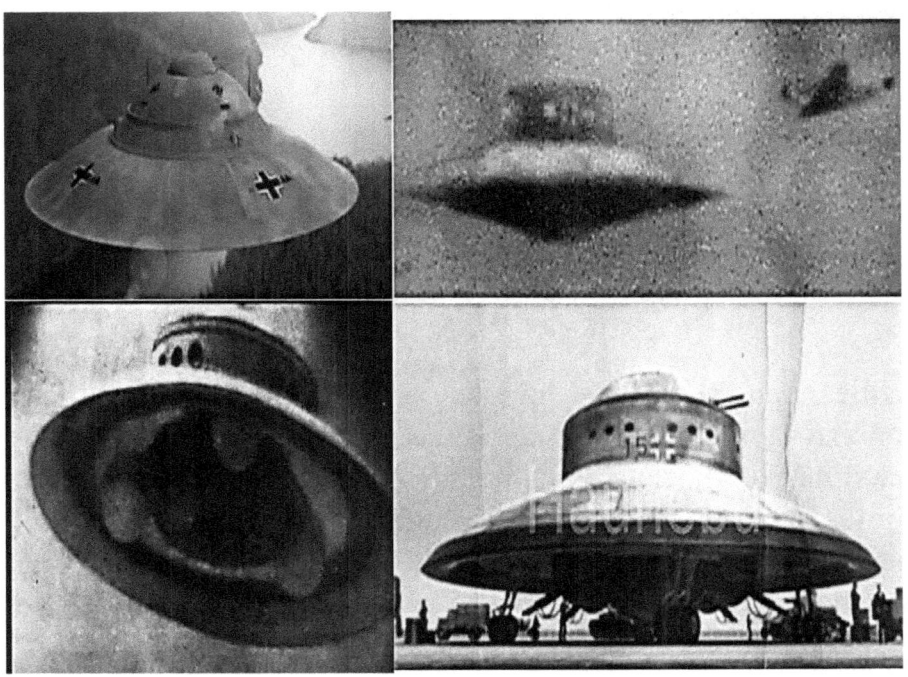

The images following are some of the declassified Flying Saucers of the United States. These were made in the mid-1950s up until about 1990. These 5 make up the main flying saucer designs that were attempted. While it is believed their flying capability never rose to the level of space flight, we do not know if classified programs went beyond these first programs. While most are 1 or 2-person machines, notice the 40+ foot wingspan one upper left. Some indications that this has been photoshopped is possible.

While our redesigns of ancient flying machines are interesting, we cannot imagine the magnificence of the flying machines of the late Pleistocene and Early Holocene Ages. Some of the hundreds of images and shrines to flying machines are shown below. As this book is not about Flying Anomalies covered in a separate anomaly book, I am providing only a small sample here.

The ones above are mostly from India where they called flying machines Vimana. Below are a couple of hundreds of those from the Sumerians, Babylonians, and Assyrians. These are typically called Merkaba. Some are shown carrying several passengers and being "charged" by hovering over a strange filamented device.

From data provided in early texts, early man-made trips to the Moon and nearby planets. Once they go there, the question might be, "Did they colonize the areas?". Let's check out the moon.

Lunar Colonization

Lunar Building Anomaly

What if there are the remains of buildings on the moon? What if there are signs they are not all "remains" but may have active tenants?

I know you're probably having a hard time with this whole war on other planets thing, but look at the evidence with an open mind so what if I were to tell you there were buildings and ranches. Now that there is interest in finding the truth, people are finding evidence of life on the Moon as outputs must have been formed to support colonization and probably even war.

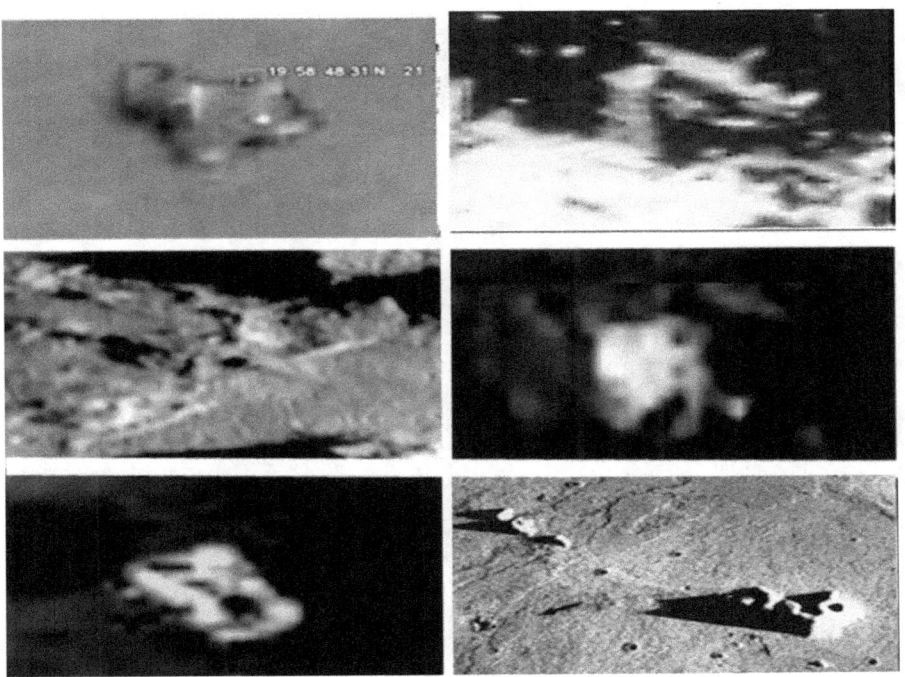

The first image is especially interesting in that 2 light sources seem to shine on occasion. As you might expect the Moon may also have played an important role in the Space Wars that must have climaxed when Venus was destroyed. While there are no major signs that it ever had an atmosphere, it possibly was a staging spot for exploration and a planetary supply outpost in the good old days. The next collage shows more of the same as we find evidence of ancient settlements all over the place.

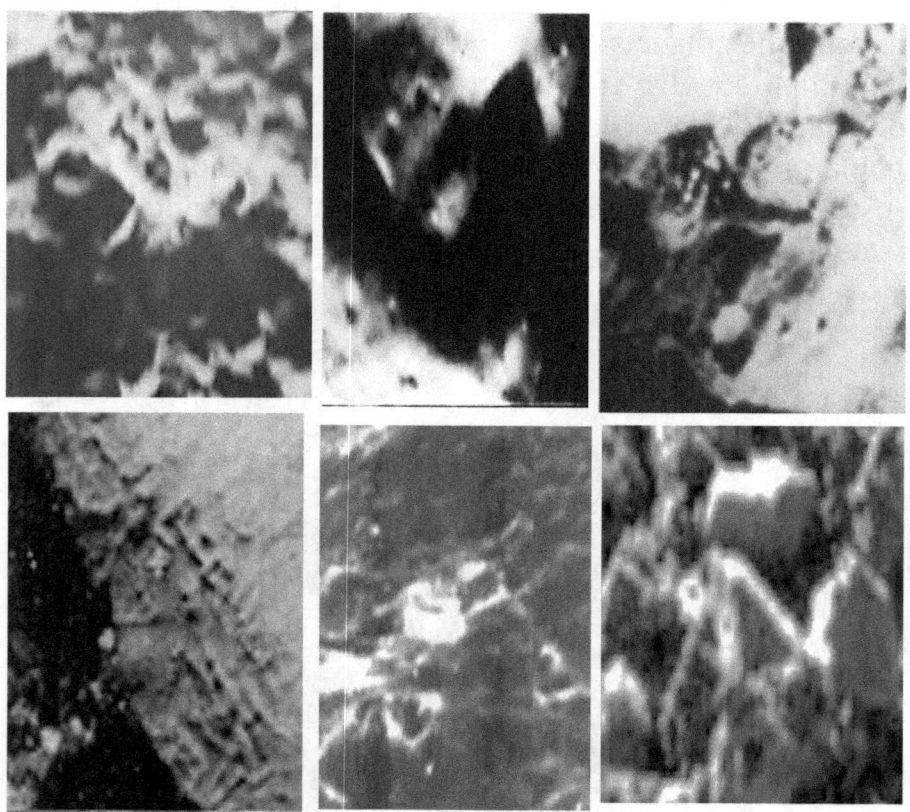

The next collage shows a group of buildings with one of the largest signs I have ever seen [upper left]. We find circular buildings, buildings that appear to be skyscrapers, and arrays of lights inside craters that sometimes are switched on. We also see what appears to be a roadway to a small town.

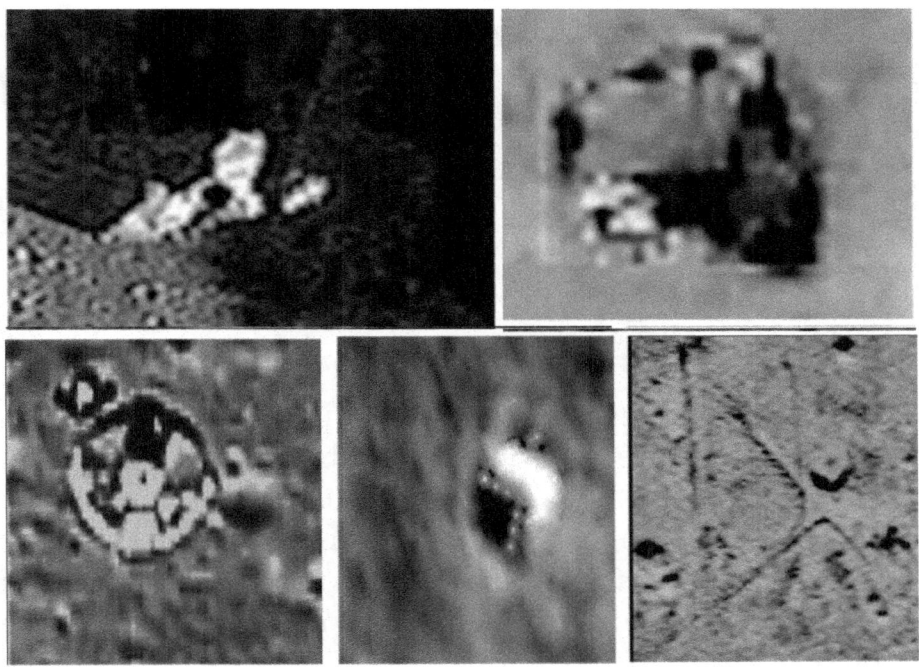

Amusement Park Anomaly

Crater Antics- On the other side of the moon we find something interesting. It looks like something is in the center of a crater. In the center of the crater we find the normal stuff. Circular depression, island shape, water-like basin and of course the amusement park in the center with a huge roller coaster and what appears to be a teeter-totter.

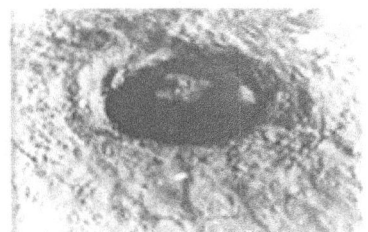 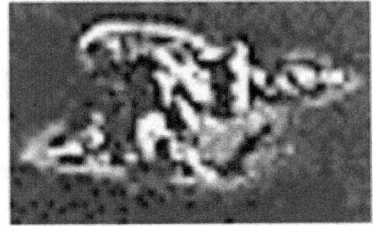

Reflecting Anomoly

Here is another odd thing that does not look natural. I call it the 2 circles, backbone, neck and one arm thing.

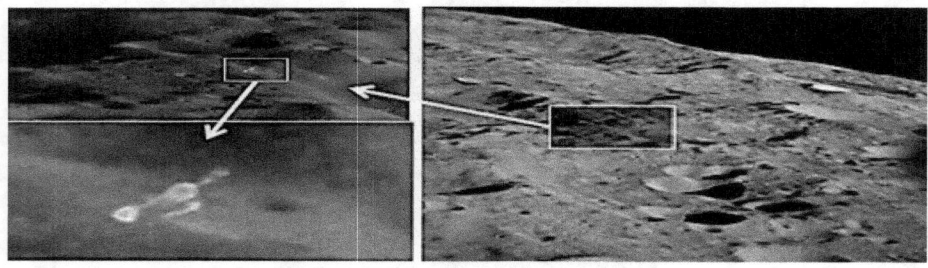

Strange Objects Anomaly

The following collage is very weird with square building on the top row, followed by some type of bridge, a building with what appears to be 3 wagon-wheels making a fence, then a peculiar crater with smooth sloping inside wall like a radar cone. The last row shows what appears to be blast holes with each hole the same size, circular, and evenly spaced. In the middle, we find a billboard that looks like a face.

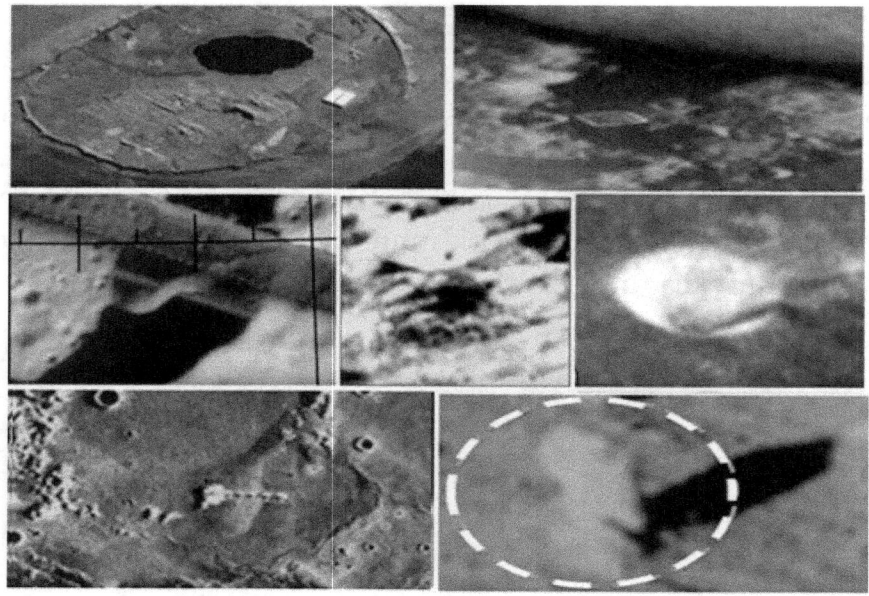

Four Mall Anomaly-Then we come to malls, or what appear to be Malls. These strange objects are all over and don't appear to be natural, but I don't know what they are. Four of them are shown below. I have tried to draw them in more of a 3D fashion below the photographs, but it is not known what they really are.

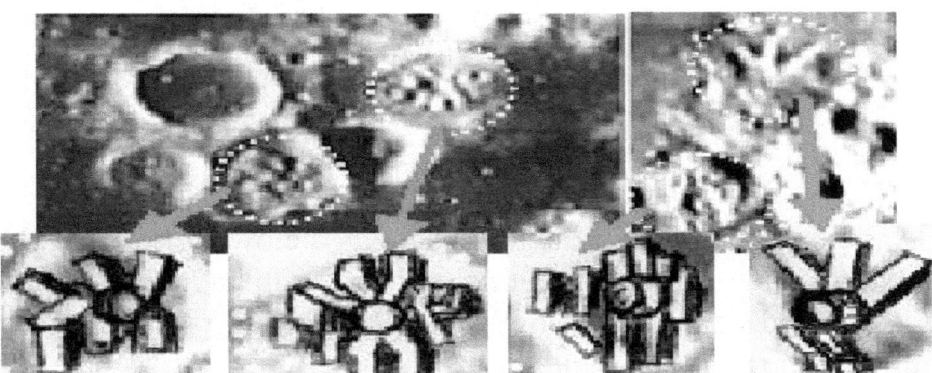

How about rectangles? A rectangular object in the bottom of a crater with a long "runway type object along its edge looks manmade. The rectangular object towards the bottom of the photograph completes the unusual topography Still another square structure with high walls is show right.

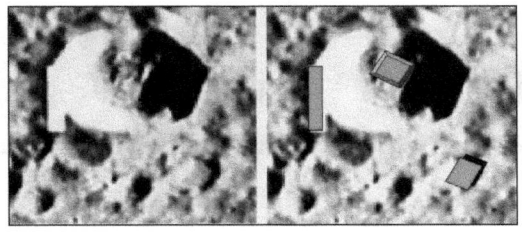 

Hex Wall and Ball-The next group provide two distinct anomalies the first is a hexagonal fortress with a rectangular building inside. From the shadow, you can tell the walls are very high. From its reflection, we can tell it has very smooth sides. Some believe this is man-made, but wouldn't someone have to be on the moon to build it? The second image is a dome. The curved structure is "unnatural".

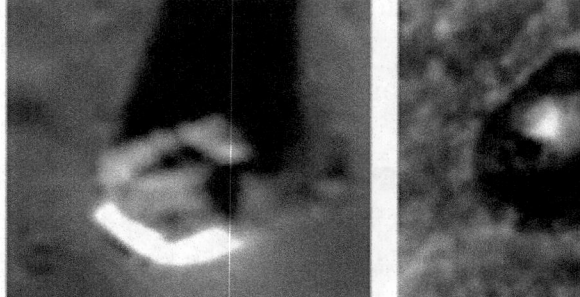

Building with Satellite dish and a Pyramid

Another strange building is shown below. The straight shadows help us make out this building with an angled object at the end of a pipe. **Long and Rectangular-**Speaking of threes, here is an image of three huge almost parallel rectangular buildings or similar structure.

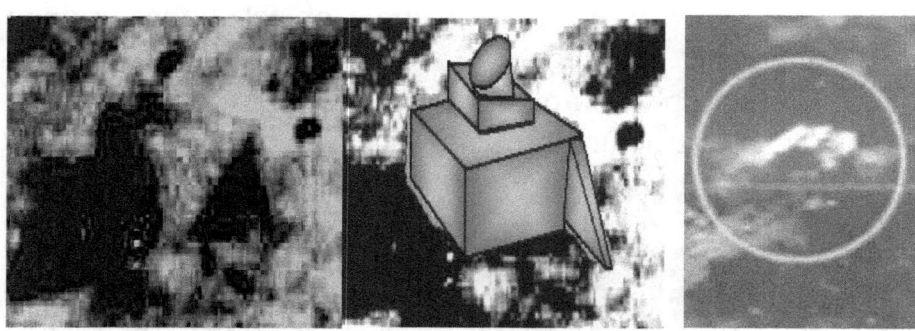

The next fortress is interesting in that there is a visible wall around the structure, that is situated on a high plateau and the building is well defined.

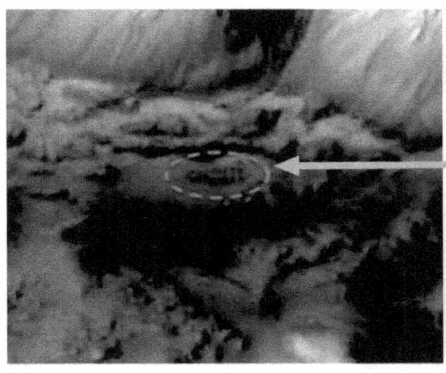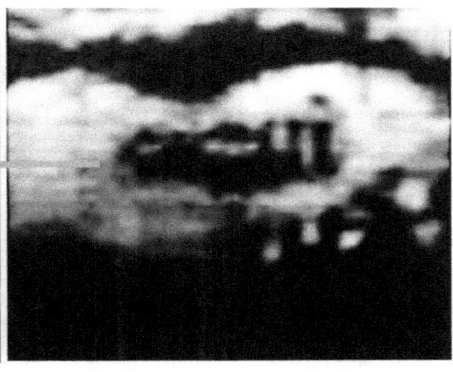

Pyramid Anomalies

The thing to the left looks like a three-stepped pyramid similar to a Maya building. I have outlined it for clarity, but there is no mistaking the regular lines and the slopping terrace and even steps. This picture is a segment from NASA image AS11-38-5564. To make this example more reasonable, another pyramid has recently been found. Most things in nature are not square as this building apparently is. Also, most natural things don't have three stepped areas. Please don't think the Maya are on the moon. That is not what I'm trying to present.

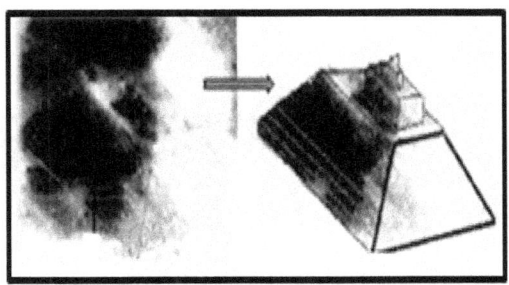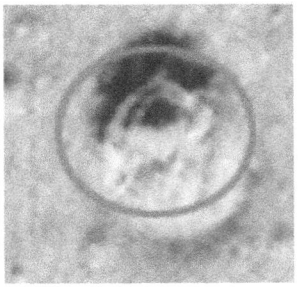

More Building Anomalies

The regular features, right angles and crisp shadows reveal the first structure in the flowing collage. Here is one that is, I believe, very plain to see. A building with right angles was built on top of a round hill and was pelted by smaller meteorites at an ancient age possibly 11 thousand years ago Next to that is a building with a tower and shapes that appear to be an E, an A, and a T. Could this have been an ancient restaurant for people who spoke English before there was an England? The last image looks like another lunar signpost pillar with a hole near the top.

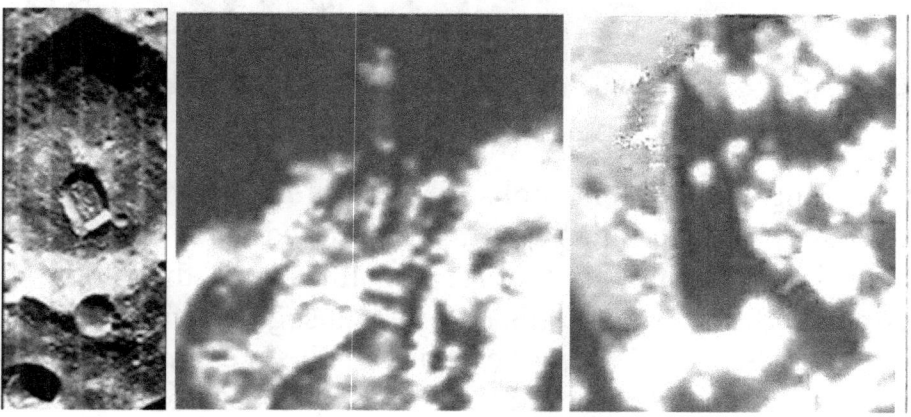

Crisscross Anomaly

The next few are more unusual object. The first are more billboards in the middle of nowhere and the second image appears to be straight-line excavations, but the last image is really strange with two multiple, bright, crisscrossed lines near one another.

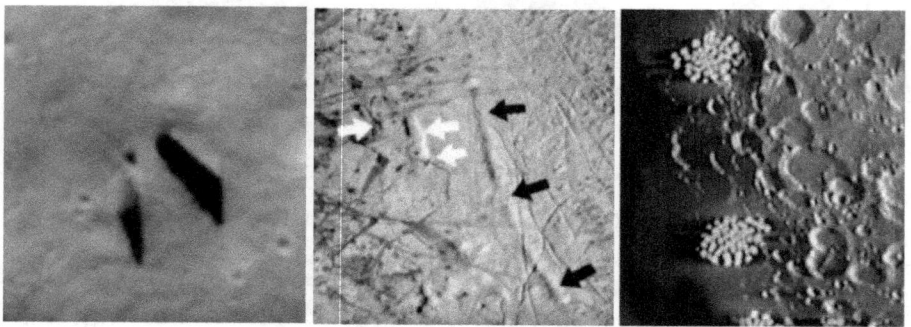

Lunar Tower Anomaly

Let's look at some of the anomalies as we begin to see the moon a little differently. On the following page are pictures taken during various flybys of the lunar surface. Look at the immense shaft rising from the surface that was seen during the Lunar Orbiter II flyby in 1967. The frame number is LO-III-84M for those who want to check it out from other sources. Some tried to explain it away as a shaft of gas erupting from the surface, others said it was an optical illusion. Look at the shadow of the illusion and note that gases would have been dispersed as it rose from the surface. This shaft is solid over 2 kilometers high. The shaft is probably man-made and taller than a 700-story building. I don't know what it is, but it's not an ordinary rock.

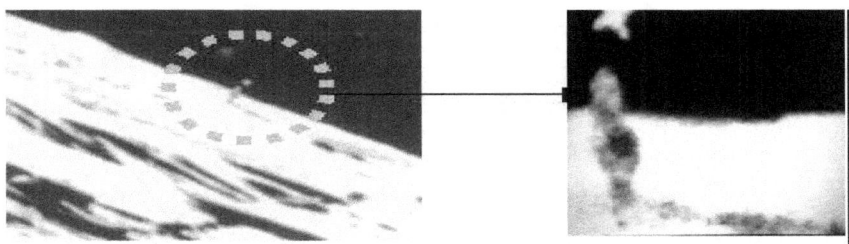

More Towers- In the following collage we find all sorts of towering structure. The first one appears to be a rocket with its pointed nose-cone. The second one is out in the middle of nowhere. We can assume to increase telephone coverage or something similar. The last one is interesting is that it is right next to one of the roadways found on the moon. Of course, it also must be manmade. The detail of this tower is simply amazing.

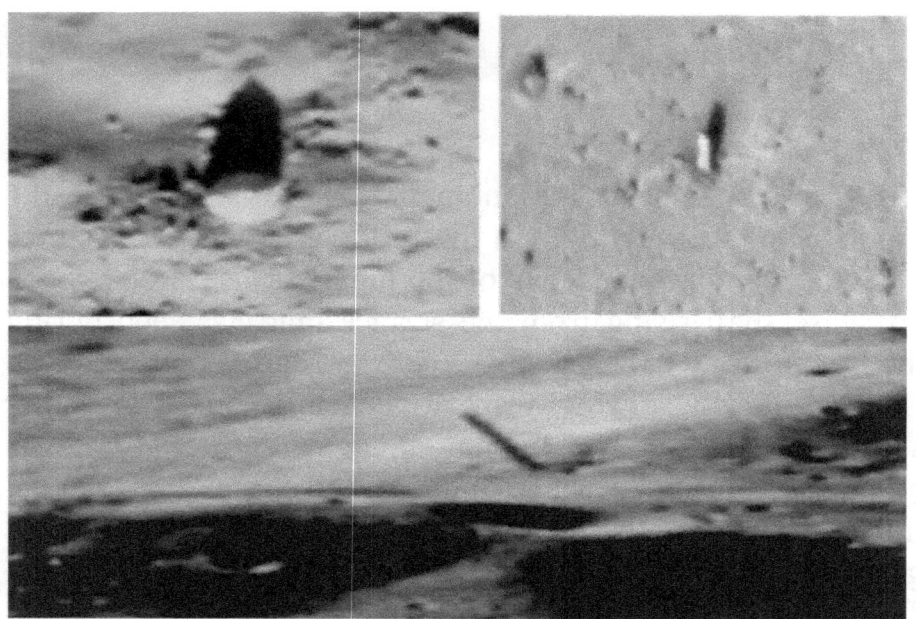

Still More Towers-Near the Sea of Tranquility, Lunar Orbiter 2 photographed this strange right-angled structure in 1966. The obelisk in its center has been determined to be 200 meters high.

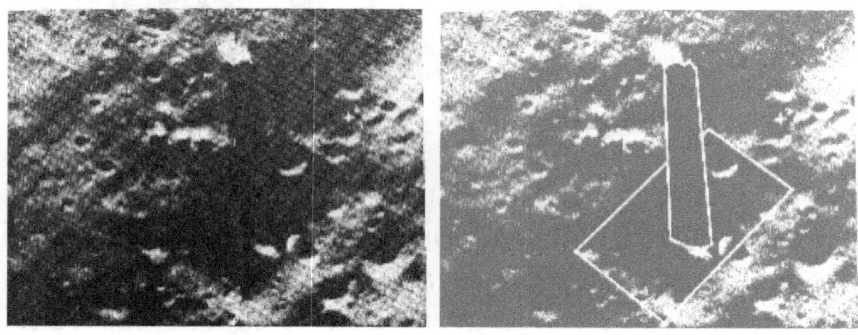

More Towers- As the picture shows, there are 7 anomalous spires constructed in the area. Although we can't see them directly, their shadows are precisely shown. These are commonly known as the Blair Cuspid's after the initial investigator. They were photographed in 1966 by Lunar

Orbiter III. The Lunascan Project has reintroduced this photograph number, LO-2-61H3, as a set of unnatural objects to be investigated further. Maybe the investigation will show us friendly neighbors. [After thousands of years of resupply and investigation on their part without harm to earth bound humans, it is obvious that the "visitors" do not mean to harm us.]

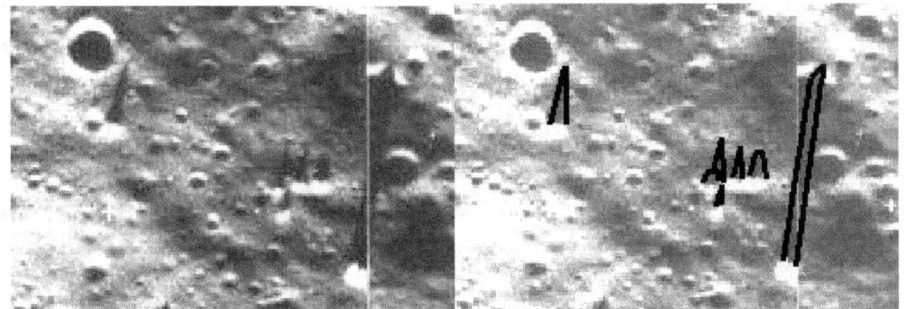

Towers in Craters Anomaly

Don't get me wrong about this group of images. The center of craters can often have debris billowing out of the middle, but these are very strange in a number of ways, the centers seem very conical and the towers appear to be flat-topped. For signaling and better vision, the inhabitants might need towers from which to view their landscape. The pictures below seem to be towers that are evidently all over the lunar surface, so they may be areas that will eventually show us other signs of "visitors". Many of the craters on the moon have an upwelling in the center as material is sort of splashed around due to the low gravity, but I have outlined three "upwellings" that don't seem to be right because they are simply too tall, very conical in shape and have flat tops. I have circles these towers to bring them out a little, but they are not very hidden even without the outlines. Also notice how round the bases are. They are more round and smoother than most craters. Typically, a crater would have jagged edges along its rim and many of these structures are perfectly smooth as far as I can tell.

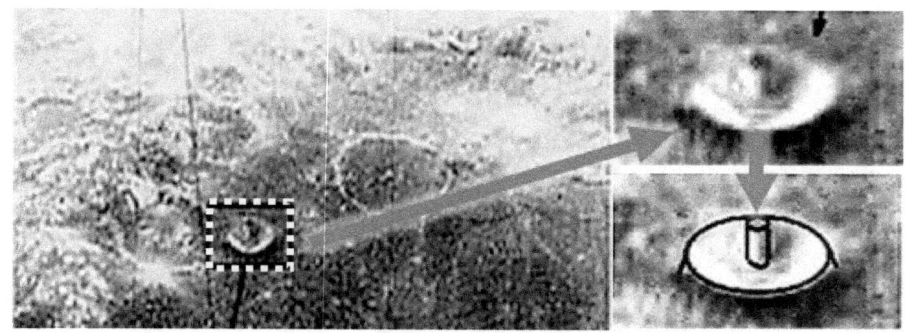

If someone was or is stationed there, under the lunar surface, the people would certainly have to come down here once in a while to restock goods and people would see them from time to time. That would be a reason to hear about the thousands of sightings of unidentified objects and unusual or unexplainable experiences. There are too many sightings to only be for experimentation but continuously getting supplies for an inhospitable place like the moon, makes more sense. Notice also the very distinct tower shadow, the straightness of the tower wall and the flat top. None of these would suggest a standard upwelling in a crater center. Still another "upwelling shows a remarkably straight tower-like structure and even straight high walls around the periphery.

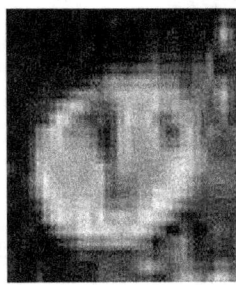  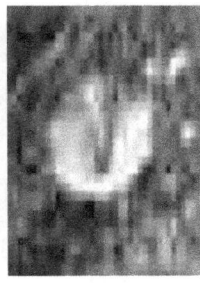 

While it doesn't make sense for people to live on the surface of the moon if there still are colonists here, on occasion, someone might have to surface to clear up debris and whatever. What if we caught some of these actions? Would that make my original statements more palatable?

Machinery Anomalies

Construction Work Anomaly-If you need construction work done, this massive thing on the moon might be able to help and it certainly looks like activity to me. It almost looks like an enormous forklift, doesn't it?

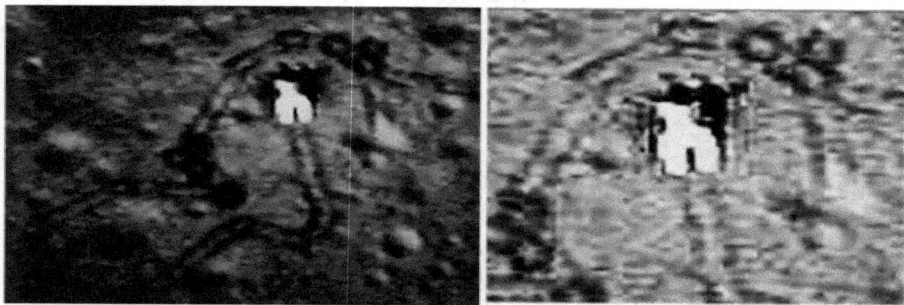

Moving Rover- Scientist put a rover on Mars, but there should not be one on the moon. Here is another thing tearing up the place was photographed below.

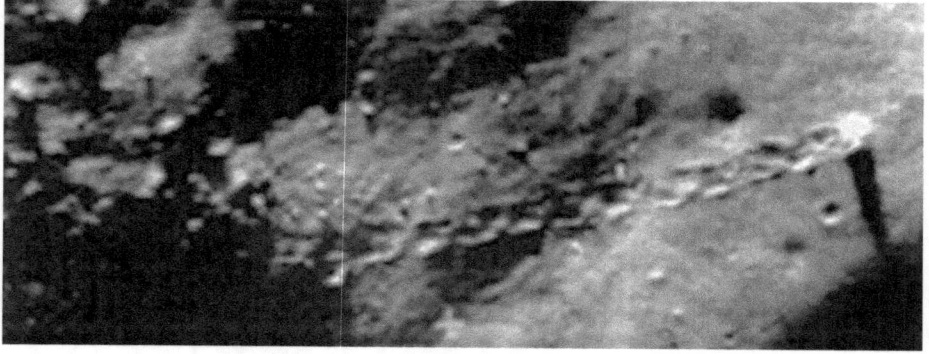

Rocket Nose-cone Anomaly-What if you could find a nosecone of a rocket. Then would you believe someone has been on the Moon at some time. What is you found a second one that looked a little like a massive ketchup bottle?

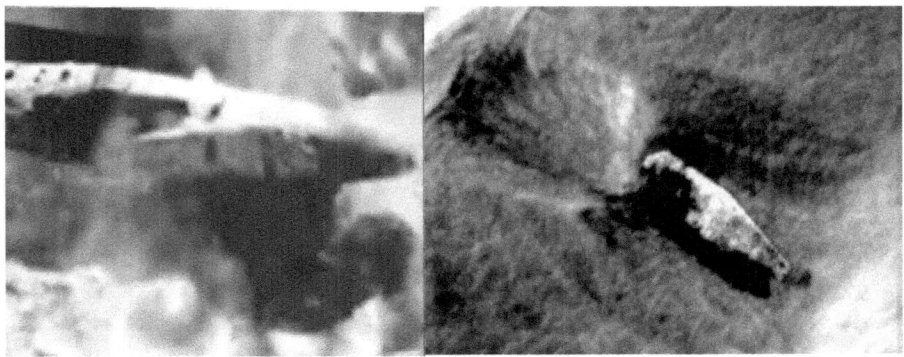

Plume of gas- The next image might be someone barbequeing and things got out of hand making this massive tower of gas thousands of meters high. Maybe something more intense is happening under the surface.

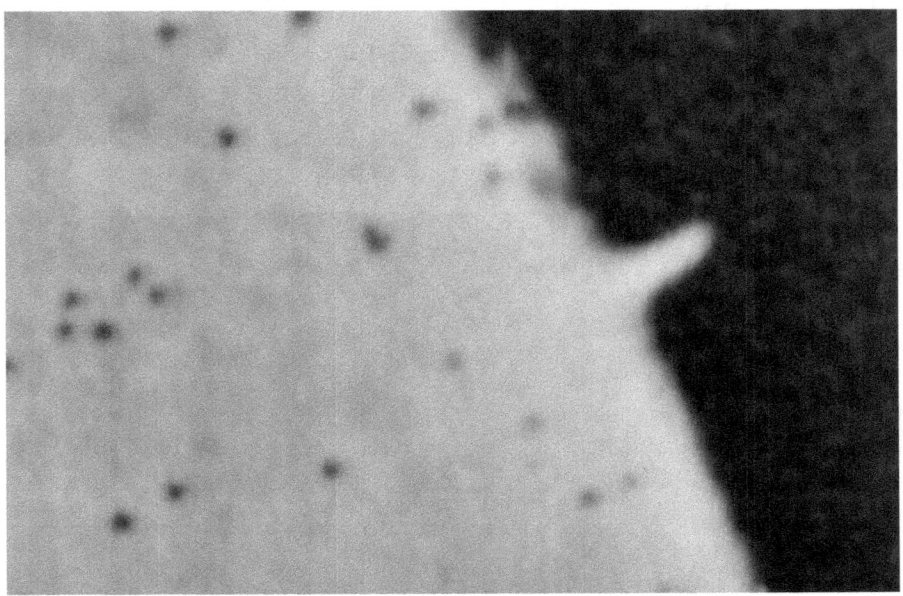

Bright Light Anomaly- We don't know what was making this light, but it had to be bright and just outside the range of the following photograph, but they must be very bringht.

What if it was a flying scaucer shapped vehicle like the "possible" one to the right in the following collage?

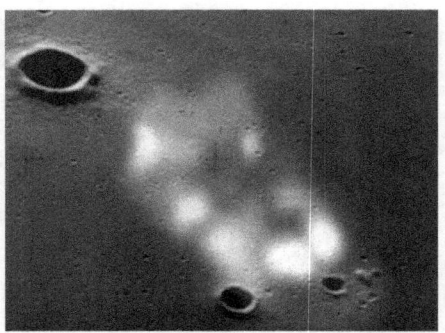 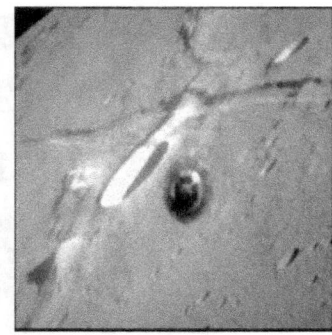

Flying Anomaly

If apparent flying saucers and rocket nosecones are being found, do they fly around the place? The Apollo mission didn't help the confusion. The following picture clearly shows a flying object moving across the lunar surface. Please notice the shadow on the surface. To the right shows a low flying something and the bottom image appears to show a low flying circular object near a trapezoidal, domed building with massive doorways.

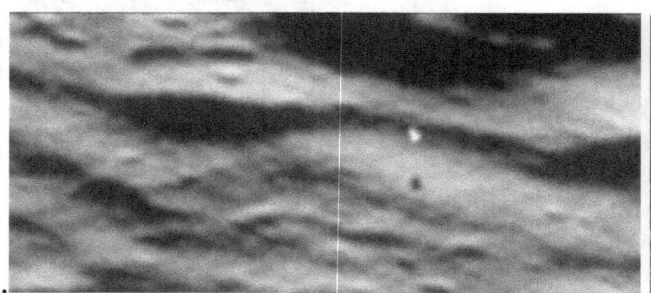 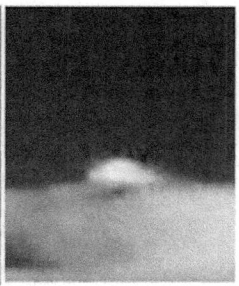

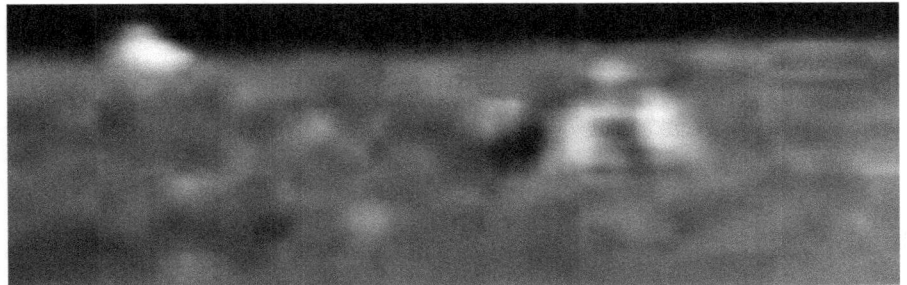

The followihng two examples are more of the same as one mission captured what is believed to be a non-earth initiated sattelite around lunar and the last image shows a timy dot that moved across the landscape as if flying.

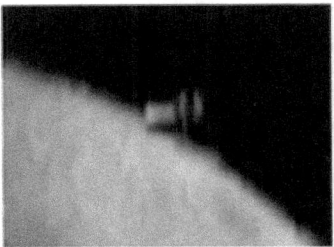 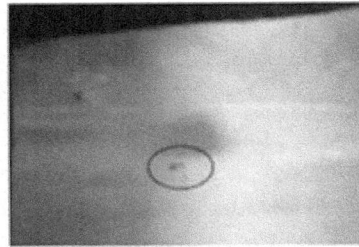

The purpose for showing you these images is not to provide proof of aliens flying around the moon, but allowing you to decide from available images. So, what about roads?

Lunar Roads Anomaly

Not only are there building, walls, billboards, amusement parks, towers, radars, pyramids, construction equipment, flying machines, and strange lights, but also there are extremely long-distance, straight lines or roads in the sand, so we can believe those who lived or live on the moon may also sometime have taken land vehicles. Here are a few of the images found.

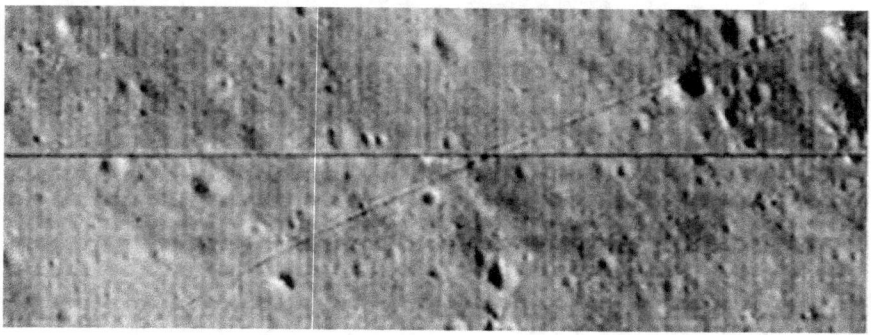

The next one is very interesting in that you can see it is a two-way structure with parallel roadways.

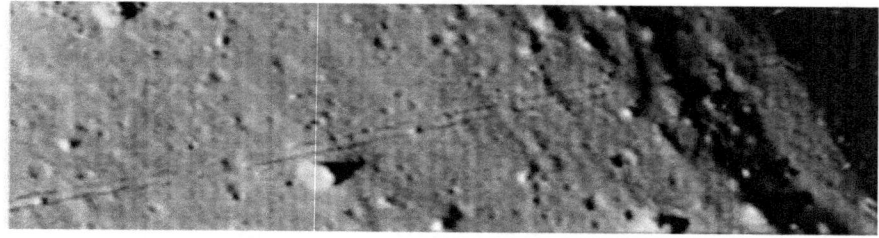

Rolling Uphill-David Hatcher Childress, in his book "Extraterrestrial Archaeology", showed NASA photos of moving objects on the surface of the moon. I don't mean moving down a hill. I mean moving up steep crater walls and down the other side. All of these movements were done without the presence of wind. In fact, there is no atmosphere at all. NASA calls them rolling rocks. Maybe they should call them "uphill rolling rocks" so that no one gets the wrong impression. The picture below shows a long, long distance traveling line that parallels another. Up and over rocks and things, I'm sure it's a natural phenomenon or else it would be an anomalous road like the others.

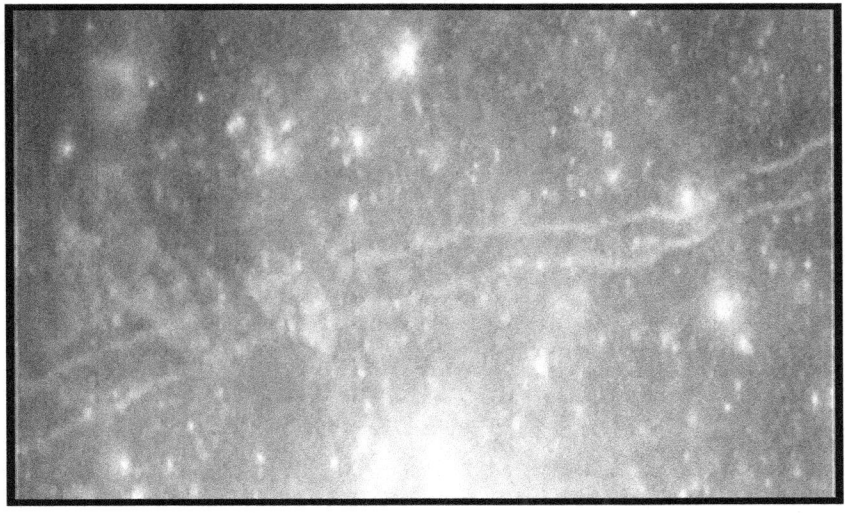

Walking Anomaly

If the humans once lived on the moon and went around amusement parks, we might expect to find holes. If we look at another Clementine lunar image, on the following page, LO-III-123-H1, we find something very strange that has never been defined. It is shown next. The surface around this area is covered with strange double-hole patterns. No one knows what they are, and they have not been satisfactorily identified as natural. Maybe these things have something to do with those trying to live on the moon.

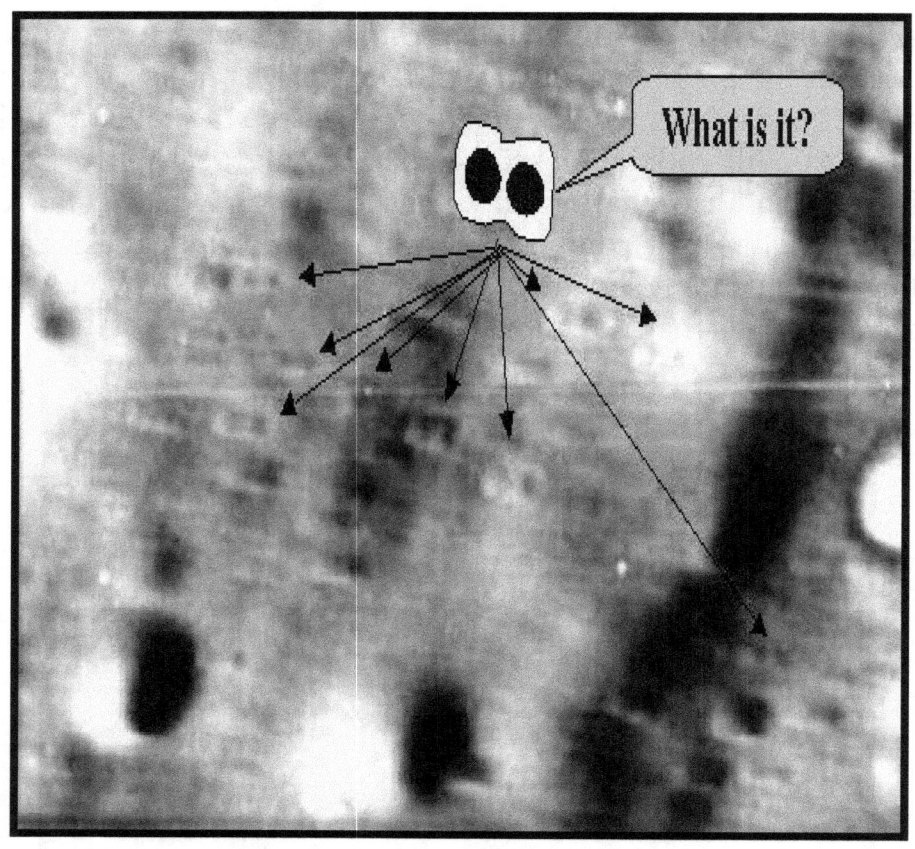

The patterns show that whatever was "walking" went in a walking pattern and shifted its weight from back to front just like we do when we walk, however, these "footprints" are much too large to be from humans. They would have been made from some type of walking machine. You can see that they are all over the place, so whatever was walking did it a lot. From huge, let's go to tiny.

Microbe Anomaly

Here is some interesting data. Scientists have now concluded that the jumble of circles to the left and right prove life existed on the moon at one time. The structure is identical to fossilized microbes found here on Earth. Microbes don't just appear something living has been on the Moon at one time.

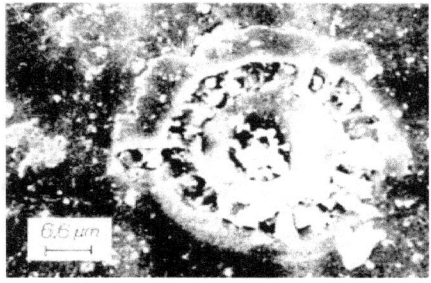 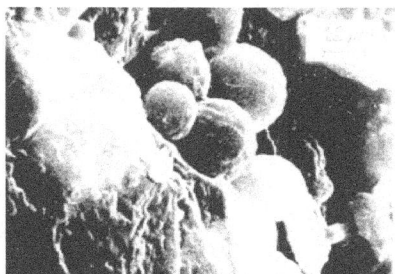

Rivers

If we find riverbeds, that may mean water flowed on the lunar surface at one time. While we have found water at the poles, there is evidence to suggest, there was an abundance of water for some level of colonization.

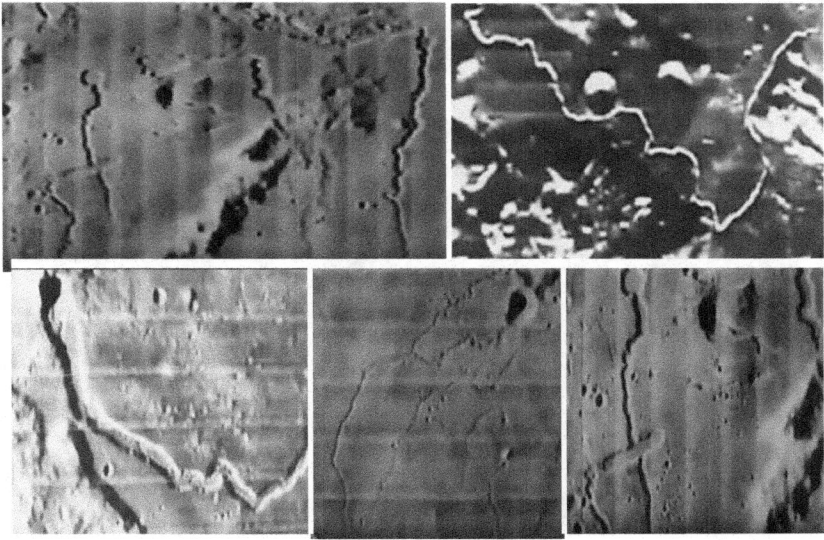

I know the previous items have not been convincing----- Wait a minute! -- They have been convincing, but there is more, so let's continue.

Lunar Blast Anomaly

There is another group of features that are very strange on the moon. Those are long straight sequences of blast holes. Some have tried to explain them away by saying a rock hit the surface and kept on bouncing for tens of miles. Although the rock hop thing sounds ridiculous, another alternative is a bombing run and no one wants to say that so the hoping rock stays on top. Just imaging a huge meteoric rock bouncing as many as 26 times before finally landing and then it gets even more bizarre as the high bouncing rock simply disappears.

Bubbles-Still another sequence of what appears to be bubbles along a fracture line is unlike all other fractures on the moon, but I don't know how it figures into habitation. It is just data, but that brings us to noses.

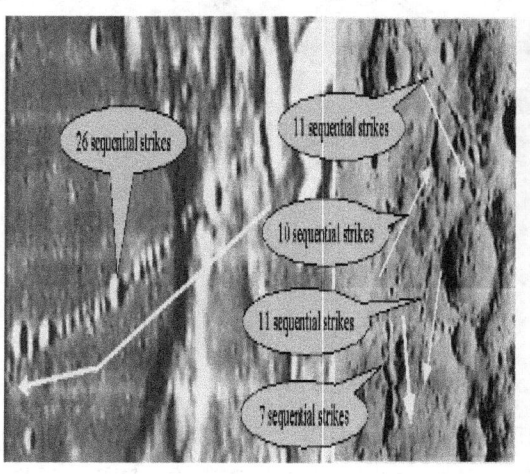
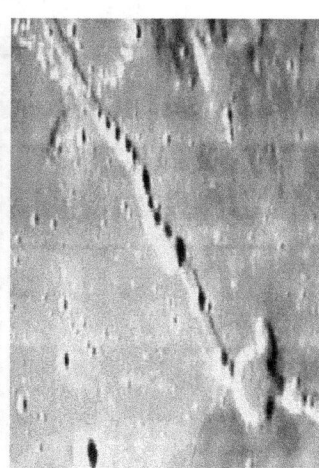

Tiny Nose Anomaly

Certainly, pilots of UFOs could not live on the moon. They look too alien. The problem with the "normal" suggestion is that the images we have seen <u>are consistent with those living on either the moon or Mars</u>.

Tiny "Alien" Noses Show They Live Underground

The skull on the left below is reported to be that of a "UFO pilot" that subsequently lost his life. Like many descriptions from eye-witnesses, the bone structure matches- large almond eyes, small nose and mouth, large skull, and generally smaller than "normal" humans. They are also said to look more delicate as well. If we assume that these pilots are human, it seems reasonable to assume that wherever they are, there is plenty of oxygen [tiny nose], little light [large eyes] and low gravity [small muscles]. Does that sound like the moon?

 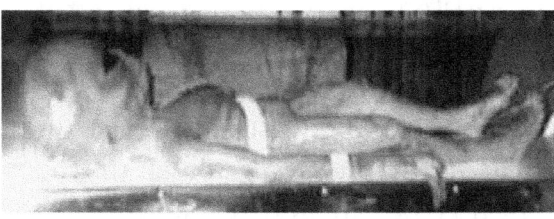

Having low gravity fits a life on the moon and if the people were forced to live below its surface, having low light might be reasonable, as well, so what about the overabundance of oxygen. For that we need to look at sand. The moon has an overabundance of sand [silicon dioxide which is mostly oxygen], so to say it has no oxygen is a misnomer. All that someone has to do is convert the sand into oxygen and silicon and people start breathing. Manufacture of oxygen in this way may be fairly efficient, but on earth our air is mostly Nitrogen. Eighty percent of our air is not oxygen so we have to take in 5 times as much air as one would take in if the atmosphere was "MOSTLY OXYGEN". Given similar air pressures, our noses and lung capacity could be one-fifth the size that it is today. We would begin to look like our [human] visitors if we made our own air. We would begin to look like the, supposed, pilot shown above. When we see pictures today, they may or may not have been doctored. We can never be sure of that these days, but it would make sense to look that way if the pilots were HUMAN and lived under the surface of the moon and made their own air. Oh! I forgot the larger brain, which shows that these humans have been stranded since before the Bharata War.

Quick review of the Bharata War Destruction

For those not familiar with the Bharata War, let me give you a quick overview. While we are not exactly sure when the war started we know it ended around 3100BC as recorded in many works from the Inca, Maya, Ughu Mongulala, Egyptians, and Indians. The end date has been verified by many scientists. The interesting thing about this particular war is ½ of all major human DNA mutations occurred during this war as if massive nuclear discharges affected everyone. Our brains began getting smaller after the war and is now 20% smaller than the Cro-Magnons and smaller than that of Neanderthal, not because they were more evolved than us, but that we have devolved from them because of this war. One thing to note here is that humans had the capacity to fly and colonize planets before the war, but afterwards, only having the ability to use about 10% of our brains also made us forget many capabilities from our ancient times. Melted cities and human remains that are still radioactive make up most of the physical evidence of this horror, but some people must have been isolated during whatever caused our massive mutations. Possibly they were living on Mars or Lunar and they heard about the destruction over radios---who knows--, but it appears these survivors have been testing the planet to see if they can resettle here without instigating another event.

I know some of this is sounding like a stretch, but what I have been trying to stress in this whole set of books on ANOMALIES is that if most of the pieces gathered from many areas seem to be painting a similar picture, then even the more improbable becomes the most likely if the pieces fit. Rather than simply ignoring 90% of the evidence to assure the 10% can be easily defined, maybe there is a theory that can come from all of the evidence. For instance, if people did live on the moon and Mars, would there be thousands of sightings as we have seen since the time of God incarnates return and possible before that time?

People Coming Back

If the "bacteria" or whatever caused our mutations is still around, individuals coming back might have the same genetic mishap, so they would want to test the environment over and over to see when they could return to their homeland. See if the following items don't sound familiar.

- They would run tests on people and animals and be seen by people but they would leave without substantial interaction.
- Sometimes they would crash and sometimes be found dead.

- The machine technology would be beyond the capability of people with bacteria infected DNA.
- People affected by the bacteria would not be able to communicate telepathically any longer, but unaffected people would still have the capability.
- The unaffected people would stay isolated so that they would not become infected.
- Large amounts of skin would be taken in sample tests because the skin would have most contact with any airborne contaminate and would be of special interest.
- Sex organs would have been tested as the easiest area to determine chromosome and DNA modification.
- The unaffected people would continuously come up with experiments to see if they could devise a cure.
- These experiments would have to include humans as the only true test of a cure or disease eradication monitor.
- The unaffected people would use representative creatures such as cows [cow chromosome similarity is so close, scientists are attempting to make the animal produce human blood.] for any destructive testing do to moralistic views.
- As part of this testing they would certainly take blood as a transport material that has contact with most parts of the body.
- The unaffected people would, most likely, see if sexual interaction could breed it out.
- As part of a breeding experimentation, the fetus would, most likely, be taken quickly rather than allowing the host

to produce offspring. Like us they would not care to experiment on a fully developed baby do to moralistic views.

- The experimentation would follow control groups to see if there was a change in our chemistry that could allow their reintegration back into our society.
- The unaffected people would be more advanced than we are today.
- Due to our common ancestry, the people would have the same defect with respect to getting into heaven as we currently do and "God incarnate [Jesus]" providing resolution would have to be identical.
- The people would be physically similar but not exactly the same as us after the 5 thousand years of "separation".
- Because they were still able to use it, their brain size might be larger.
- Reasonably long periods in weightless areas might cause minor physical differences like less muscle material or larger eyes if their living areas were darker, but they would remain humanoid and recognizable, because they would not be a different kind of creature.

Guess What!

All of these items go along with the "pilots" we have been seeing and attributing to UFOs, but where do they go and how to they keep from being infected? That's where a hollow moon comes in.----stop laughing, right now or I won't tell you the findings. OK! I'll still tell you, but my feelings are still hurt.

Hollow Moon Anomaly

I know the idea of a hollow moon sounds bizarre and I don't mean an empty shell, but rather there is a possibility that there are fairly large underground pockets directly beneath its surface. It sounded absurd to me when I first heard it, but if you wonder how the moon could have a density of only 3.3 grams/cm^2 instead of the earth's 5.5 grams/cm^2, like Earth, you're not alone. <u>If you wondered why the moon's revolution and the earth's rotation are "exactly" identical, you again have company</u>. Here are some expansions of the questions and explanations. The most likely way for the specific gravity to be so low is for the moon to be mostly liquid or pocketed. The most likely way for the moon revolution and earth rotation to be identical is for both the earth and moon to have once been the same unit. If they are both pieces of the same object, it is reasonable to believe that their densities would be similar.

As I went over earlier, new data tells us that the moon came from the Pacific Ocean when the Ocean was scooped out in the first place. What I want to investigate is what makes the moon so light? The first thing we might notice is that the earth's core did not separate whenever the moon was formed. Only the lighter crustal portion was sent into space, but that, by itself, does not explain the extremely low lunar density. Some possibilities of what makes the moon density low are given below, but only the last two deserve any reasonable consideration even though they also seem rather bizarre.

The moon is not a natural thing. *Some have presented the hollow problem means the openings were not natural, but the immense size keeps that possibility extremely low.*

The calculated density measurements are in error*, but the measurements were determined from known values of gravity, moon dimension and orbital distance, so that seems unlikely.*

The moon is made out of green cheese*, and space mice could have eaten holes in the interior of the structure. Silly but sometimes silly is not wrong. In this case, it's just silly because the moon is certainly not green.*

Portions of the moon were excavated in the early days*. This one makes some sense in that the mineral content of the moon rocks brought back to earth is extremely low in useful exotics. One might wonder, "Why would the content be different than that of the Earth if the moon came from the earth?" One strong possibility is that the minerals were removed.*

A hollowed-out area causing the low density was produced by some highly intelligent human group- *This would be done if the group wished to live close to the earth but not so close that some form of bacteria could modify their DNA.*

I know the moon doesn't appear to be a mineral excavation site, nor does it appear to be a hollowed-out housing development for humanoids. The Babel discussion presented previously should get you thinking about how the truth sometimes doesn't seem to instantly make sense and things that you have been taught as laws and absolutes are not always absolute. Here is the thing to remember. Things that are not absolutely wrong must have potential for being correct.

Trace Mineralization Anomaly

By the way, humans using the moon as a base of operation and living makes sense when it comes to trace minerals. If the Moon came from the earth, then the trace mineral contents would be similar. These trace minerals affect our bodies in important ways and humans would live better in an environment with similar consistencies of these important elements. Mars and even Venus could be less desirable in that regard.

Echo Evidence Points to a Hollow Place

Here is one piece of evidence that seems to confirm a hollowed-out moon. This is an important one so listen! During the 1969 landing of the Apollo 12, the sound of the Lander hitting the surface made a vibration that <u>lasted almost an hour.</u> This is well documented with the seismic equipment on board and NASA records. The sustained vibrations strongly suggest that the moon is, indeed, hollow in some areas. Go figure. It must be honeycombed with thousands of cavernous underground tracks that could be turned into oxygenated labs or living quarters if someone simply knew how to get all the oxygen out of sand.

People Underground Anomaly

I know it sounds absurd to say that people live on the moon, but if there were some people living there, the question might be, "Where do they live?" If these people live on the moon there is one thing that is almost assured. They do not live on the far side of the moon. The far side has been bombarded many times more than the near side because the near side is always protected by the earth. There is no reason to even check on that side. The near side has about 30% of its surface fairly free of craters, while the far side has less than 3% clear. The near side is a convenient side to check and even Neil Armstrong indicated that he saw spacecraft on this side of the moon during his Apollo 11 mission, which may peek our interest. What I'm trying to say is, not only have there been people on the moon in our very ancient past; but they may still be there—living underground.

Living Underground-The apparent buildings, compounds, battle affects, towns, and earthlike features above seem to indicate life on the surface, but the pilot evidence suggests that the people live underground and manufacture their own oxygen.

Tiny "Alien" Noses-Tiny noses show the 'visitors" live underground. The depictions of the UFO pilots are almost always the same. Like many descriptions from eye-witnesses, the bone structure matches- large almond eyes, small nose and mouth, large skull, and generally smaller than "normal" humans. Plenty of oxygen, low gravity, live in a somewhat darker cave these human pilots would have a small body, small lungs, tiny nose, larger eyes----AND STILL BE HUMAN. Just take some silicon Dioxide and remove the silicon to make pure oxygen.

NASA Makes Lunar Oxygen-While getting oxygen out of sand is difficult, NASA has been working on simple methods to get oxygen out of the Lunar rock that contains a wide variety of Iron oxide based compounds. With as little as 800-degree temperatures, they and have been very successful making a substantial amount of life giving oxygen which also allow rockets to fire. All lunar rock and soil contains approximately 45% oxygen, combined with metals or nonmetals to form oxides and they have found about twenty different methods for oxygen extraction on the Moon, so don't worry about breathing. You simply need a smaller nose without all of our Nitrogen in the air.

Small Nose for Oxygenated Air- Given similar air pressures, our noses and lung capacity could be one-fifth the size that it is today if we were only breathing oxygen. We would begin to look like our visitors if we made our own air. We would begin to look like the people shown on the following page. When we see pictures today of these odd looking "HUMANS", they may or may not have been doctored. We can never be sure of that these days, but it would make sense to look that way if the pilots were HUMAN and lived under the surface of the moon and made their own air.

Larger Brain from Isolation-Oh! I forgot the larger brain, which shows that these humans have been stranded for a long time---ever since the Bharata War that I talked about [some call it the Tower of Babel War] The first image was found in Tibet while the second one was drawn on a cliff in Utah. On the second row, the first comes from Mexico, the second skull is from Peru, and the last image was drawn on an Egyptian Pyramid.

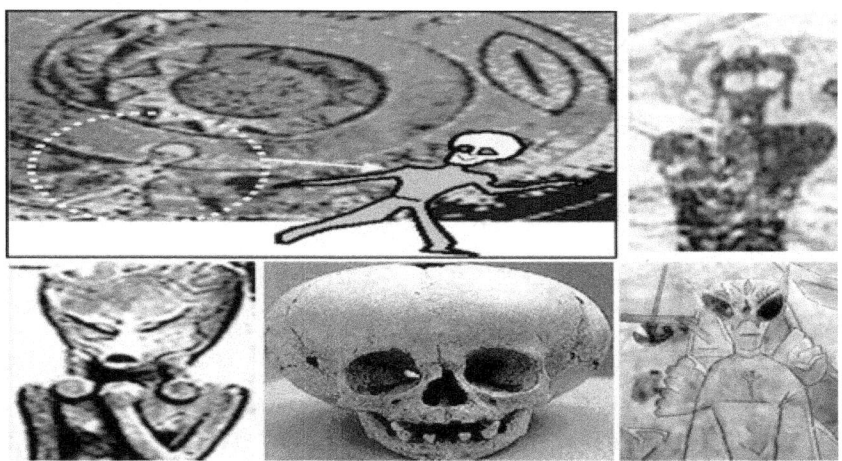

Big brain, almost no hair, big eyes, a tiny body and the nose of someone that lives where the atmosphere is not mixed with nitrogen like we have.

The Air is Confined-If you are wondering if this high level of oxygen would occur naturally, I think that the answer is no. If there were naturally forming places with high concentrations of this corrosive and volatile gas, the planet would simply explode or rust away. The high concentration most likely indicates a very small captive area such as an underground town or a domed area above ground. I know all this seems like a stretch just by seeing a small nose and tiny lungs. No matter how strange it sounds, it seems to fit better than saying any of the following.

- *"The little space beings come from a galaxy far away."*
- *"The flying saucers don't have people in them."*

- *"The UFO's don't even exist and are figments of the imagination of the thousands and thousands of people that have witnessed them over the years."*

Hopefully, the discussion above, the possibility of humans on the moon is not considered completely impossible, and the confusing mystery surrounding the unidentified flying objects is less troubling. Awareness of the UFOs has become very polarized at the present time, but even with its popularity, the textbooks are not being changed.

So, What!

What may be of more importance is what will happen to us, regardless of the UFOs, in the future. That doesn't mean we should completely forget the UFOs, the moon, and outer space, because there is some indication of a more substantial connection between the pilots of these interesting objects and the pilots may be closer than we have been lead to believe.

- *If the moon has hollow areas;*
- *if the UFOs come from the moon;*
- *if the pilots are human;*
- *and if they are desperately searching for a cure to some terrible disease;*

---it probably doesn't matter so much unless our two civilizations come together in conflict or harmony in the near future, but that is another story. Right now, let's see is Venus details have been turned into anomalies.

Venus Bible Anomaly

During the introduction, I went through some pretty strange things and some of you think it's a fantasy, so let's get into some more details just like we did with our moon.

Huge Destruction -I'm sure some are thinking the massive destruction of an entire planet like Venus is just a little too Si-Fi, so let's first review the Bible and go from there. The verses used come from Psalms, Job, Isaiah, Enoch, Revelation, and Jasher. Enoch and Jasher are not in many Bibles today, but they have been canonized and described as important works in our current Bibles. The destruction of the planet Rahab/Venus would occur during a horrible war [Bharata or Babel War] where "Jasher" indicated 1/3 of the entire population of the Earth died. We can be pretty sure 100% of the population of Venus met this same fate. Here are the specific Biblical verses that are easily interpreted.

Psalm 89:10 - *"Thou [God] hast <u>broken Rahab in pieces</u>, as one that is slain;"* [The pieces sound like meteoritic pieces. Especially as we read further.]

Isaiah 51:9- *"O arm of the LORD; awake, as in the ancient days, in the generations of old. Art, thou not he that hath <u>split Rahab</u>, and wounded the dragon?"* [The Dragon most likely was one of the ancient people or a mighty creature governed by them. Note the idea that the planet was split as will be shown in a topographical map.]

Job 26:12- *"The boastful Angel and his followers rebelled. Yahweh destroyed their dwelling places. He divideth the sea with his power, and by his discretion <u>he smashed Rahab</u>. It was reduced <u>to stones of fire</u>."* [By this verse we could well believe that many of the ancient people had made Venus their home before the disaster. To make it easy on you, I am going to interpret stones of fire. They were meteors that hit the Earth.]

Egypt Anomaly- In order to <u>**not**</u> show Rahab was Venus, Bible scholars try to show Rahab means Egypt in these verses, even when it is ridiculous. Egypt was not turned into rubble; Egypt was not smashed and reduced to stones of fire; Egypt was not around in the "Ancient Days" associated with the Isaiah verse; Egypt had no wounded dragon as was associated with the very ancient preflood times, except for the one in Persia that was destroyed by Daniel. Also, the boastful Angel, Satan, never had his followers live in Egypt for God to destroy their dwelling place. Please get that out of your head.

Enoch 85 and Revelation 9- *"I beheld a single star fell from heaven-then I beheld <u>many stars which descended</u> and projected themselves from heaven to where the first star was."* [Almost identical details are presented in both of these books as massive numbers of "STARS", or meteors, descended after the first star [Venus] fell into destruction.]

Jasher 2:5-6- *"-and the sons of men forsook the Lord all the days of Enosh [Adam's grandson] and his children; and the anger of the Lord was kindled on account of their works and abominations which they did in the Earth. And the Lord caused the waters of the river Gihon to overwhelm them, and he destroyed and consumed them<u>, and he destroyed the third part of the Earth,</u> and notwithstanding this, the sons of men* *did not turn from their evil ways--"* [I just put this here to show how horrible the war was on Earth during the Pleistocene- just before the Pleistocene Extinction 10 thousand years ago. The destruction of Venus would have been around the ending of the War about 11 thousand years ago.]

Isaiah 14:12- *How art thou fallen from heaven, O "morning star/ Heylel", son of the morning! how art thou cut down to the ground, which didst weaken the nations!* [Heylel means "morning Star" or Venus as it is the brightest star in the morning. Parts of Venus falling to the ground caused hundreds of thousands of craters and massive fires that would weakening nations. By the way; Heylel is not found anywhere else in the Bible as most accounts call the planet "the vain Planet" Rahab.

Zadspram Descriptions

If we expand our investigation to the Zoroastrian Bible, as an offshoot of Judaism adopted by the Iranians around 600 BC, we find even more interesting information about Venus. **Zadspram 5- Venus Destroyed-** *This was the first contest, that of the sky [or space] with Satan. For as he maintains the spirit of the sky, and the sky in its fortress spoke these hasty, deceitful words to God, Therefore Ohrmazd prepared another fortifying wall, that is stronger, around the sky, -And he arranged the guardian spirits and warriors around this fortifying wall, -and would not surrender the outer boundaries to Satan's Army. Satan endeavors that he may go back to his own complete darkness- to arise at the appearance of the renovation of the universe at the end of the nine thousand years. --God said he would utterly destroy the renovation of the universe; and it will stay destroyed until the end of the world, and on account of the utter depravity of the* *wicked their destruction is fully seen.* [In this scripture, it seems the "Renovation of the Universe" was colonization on Venus, and Satan was involved in the colonization. This clearly indicates the complete destruction of the colonies would never be reestablished. Satan's Army was kept from attacking the Earth so we can believe most of his army was destroyed when Venus was destroyed.]

Zadspram 6 – Worldwide Flood After the Venus Destruction-*God came secondly to the water, and let its rain fall on the night for the destruction of noxious creatures. The rain killed all the noxious creatures except the reptiles, who entered into the muddiness of the earth. The wind pertaining to the whole earth is forth, and the water in its grasp is flung out from it to the sides of the earth, and its wide-formed ocean arose therefrom. The ocean keeps one-third of this earth, and among its contents are a thousand sources and fountains, whose water is from the ocean, and all the fountains of water continually compasses and canters around them, for forty days.* [There can be little doubt this is the same 40-day rain at the end of the Pleistocene, identified in Genesis. The words "flung out", "wind pertaining the whole Earth" and Water continually compasses and canters around them" all describe the horror as the Earth axis shifted.]

Venus Colonization and Destruction

Venus Visit Anomaly

Venus simply caught on fire from greenhouse gas so the Earth is next unless we quit exhaling CO_2.

The Bible descriptions are nice as a starting point, but there isn't enough detail so we need to look farther. One area of detail might come from Global Warming. I wouldn't know where to begin concerning the outrageous claims made to push an "artificial" Earth destruction scenario dreamed up to put money in the pockets of investors of "Greenhouse" services. Most importantly, we should all know that CO_2 has a base energy transmission wavelength of 10 microns and <u>our atmosphere cannot absorb 10 microns</u>. This means, CO_2 in no way makes our atmosphere hot. There are so many things wrong with the idea of human caused global heating that are KNOWN by those pushing the purchase of windmills, solar cells, battery cars, no-Freon underarm spray, and elimination of Jet Airplanes. Carbon Dioxide emissions cannot SOMEHOW make our air hotter. That just makes me----I just vomited a little so let me go on as Venus is important to our history.

Overview

I'm going to tell you a story that you may not initially believe. As you read the words, more and more, you will see how you have not been told the truth about our history and the history of our closest planet, Venus. The story is not a fiction. It is the probable truth as derived from hundreds of bits of data found in ancient texts, physical evidence, religious dogma, and scientific research. There will be some who will not want to believe it—not because it doesn't make sense, but because it goes against what others have told you about Venus. The image following will sort of kick start our journey along with a few of the more interesting points to review.

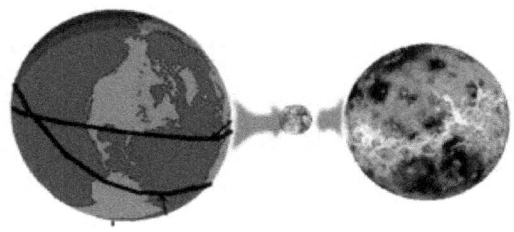

Gravity Destruction-Unlike the gravitational pulls of Mars and Earth which caused massive destruction, Earth and Venus apparently had a slightly different encounter, but they did come fairly close together. Even now the two planets are only separated by 0.05 AUs. When a comet gets 5 times that distance from us we start shaking in our boots. Before Venus was pushed closer to the sun, we can believe the distance was much closer on occasion. The preceding image show "the moon of Venus" in between Earth and Venus. This is only an example of what may have happened, but we know that all planets build up enormous electrical potentials as they spin around the Sun. If one planet gets too close to another there is opportunity for an electrical discharge called a plasma to be generated. We have a massive amount of data that tells us there was such a discharge between the Earth and Venus and the remains of that last discharge is still associated with Venus.

Venusian Tail-In mid-1997, the Soho satellite detected a plasma structure issuing from Venus and it is long enough and in the right direction to almost reach the surface of Earth. The report described the structure as "stringy." Such a structure could only remain intact if a current were continuously flowing from Venus to the surrounding space via the plasma tail. The initiator could have been uneven electrical charges between Venus and Earth or even the "Lightning Weapon" suggested in Greek Histories. No matter what initiated it, there is a high probability that pieces of Venus's moon were split away during the initial ionizing blast. These pieces would have fallen on Earth as a giant meteor storm. The discovery supports the idea that Venus assumed its present position in the solar system only recently, and has not yet achieved charge-equilibrium with its environment. When I say recently here I mean less than 40 thousand years. It also gives evidence to the probability that the reactive partner in the production of the plasma was Earth. It does another important thing as well. It makes the ancient descriptions of Venus even more believable as the planet would have looked like a huge comet in ancient times when the plasma trail was at its greatest size. It would have been a "wavy haired" planet and substantially more visible that it is today.

Venusian Heat -Today, Venus has a surface temperature of 900 - 1000 degrees F. and scientists are trying, unsuccessfully, to explain the extremely high temperatures away by a "greenhouse theory" that doesn't work. The planetary surface is so new that even the mainstream scientists are now having to devise a "global resurfacing event" [like the one presented] to explain it." We should look at all the similar myths and legends around the world describing a world-destroying catastrophe with Venus as causal agent and open up to the possibility that this well documented event could have caused havoc on Venus. Besides its lack of charge-equilibrium, Venus is totally out of "thermal balance" according to all direct observations.

Backward Spin -In order to gain some sense about its thermal problems we need to not only look at the ancient histories. or the thousands of meteoric indentions, or even the in-line blast marks on the Venusian surface. Let's look at the fact that Venus spins backwards. This would not have been primordial as the planetary spins would have taken on the general forces that were found during the solar system beginnings. Most of the planets follow this rule, but the Venusian spin was changed by some traumatic event much later. The most logical event maker would have been interaction of a close planet [possibly earth].

Venus Craters -Besides the fact that there are very few craters on the surface of Venus which shows that the surface is very "young", we come to another curiosity. Almost all the large craters can be found between 78 and 85 degrees in Latitude, but they can be found all the way around the planet along this central hub. This ring of craters tells us that the meteors came from something orbiting very close to the planet around its equator. I'll tell you what I think was orbiting Venus, but you have to promise not to tell anyone else, because they will think you are nuts. I think it was a moon.

New Volcanos- Another thing can also be noticed. Although there are many volcanoes, none seem to be active at the present time and the volcanic areas are equally distributed around the entire planet rather than being grouped along surface plates as is found on the Earth. These were very disturbing to scientists who tried to blame the "greenhouse" thing on these volcanoes. It was as if the entire planet erupted all at once.

Argon Evidence-The main curiosity, however, found by the Magellan probe was that the atmosphere contains high levels of the isotopes of argon, neon and noble gases. These high concentrations of noble gases could only mean that the current atmosphere of Venus is extremely young, because noble gases don't combine with other materials and escape easily into space; even with a thick atmosphere.

I know some of this is new to you as the details have been kept away from you to protect the "Unsubstantiated Theories".

Unsubstantiated Greenhouse Belief

Most of us, without question, believed the following unsubstantiated story. You were told that the delicate atmosphere of Venus was destroyed by a greenhouse effect that is threatening the Earth. This is preposterous, but you have been told so many times that it has sunk in.

Unsubstantiated Ancient Fire on Venus Belief

Here is another notion that keeps coming up. You have been told that Venus has been in a state of flames for millions of years, but the evidence all points to a very recent occurrence that changed Venus forever. You will find out that it happened only 11 thousand years ago.

Unsubstantiated Stability of the Solar System

You have been told that the solar system is stable and that comforts you, but the facts may not be as comfortable as this contributed to Venus going supernova.

Unsubstantiated Mesozoic destruction of Venus

Their real issue some of the quasi-historians have is that they cannot force themselves to believe civilized people have been living on Earth form hundreds of thousands of years. If Colonization occurred, they would have to rewrite their whole theory and that would be embarrassing. It is better to simply lie.

Unsubstantiated Idea Venus had no Colonists

Venus is too hot, the moon has no resources, Mars is too cold and it's a dead planet. On the surface, there are bits of truth in those claims and they certainly would eliminate colonization to any significant level, but when we weed through the surface, we find a substantially different view.

Erroneous thought that UFOs are from Sirius

We are told that the UFOs we see today, those described in ancient texts and in 15th century paintings, those written about in prehistoric descriptions, and those described in etchings as far back as Neanderthal times come from some distant star where everyone travels at faster than light speed and live for millions of years so that they can travel to our Earth because it is so interesting to the remote viewers. Even on the surface of it that idea sounds stupid, but it allows for an answer to some anomalies, so I thank those investigating this possibility because it is better than simply hiding your head.

Niberu Home- By the way, one researcher, Zecharia Sitchin, indicated that instead of invaders coming from distant galaxies, they came from a place "sort of" identified by the Sumerians. It is a planet called Niberu. This is a massive planet in our own Solar System out beyond Pluto. While

Astronomers are investigating distant indications of planets, they have not found Niberu as of yet. The reason I am bringing this up is that if there is such a place, all of that time travel and other stuff could actually work for visitors coming in UFOs. My feeling today is that it is not out there.

How Bad Is Venus?

It has almost no rotation. In fact, Venus is currently rotating about 1/100th as fast as its sister planet Earth. The thing we should concentrate on is Venus before all this happened. I know you were originally told that Venus has been this way during the entire time mankind has been on the Earth, but it is not the truth.

Today we are pretty sure Venus was a livable planet until something horrible happened around the end of the Pleistocene Age. While Venus would have had similar temperatures, water, and atmosphere, it may have been even more beautiful than Earth in some ways. On the planet are six mountainous regions that make up about one-third of the Venusian surface. One mountain range, called Maxwell, is over 500 miles long and reaches up to some 7-miles high, making it the highest feature on the planet. It's kind of like the Himalayas only longer and higher as Everest is only a little over 5-miles high. While Venus, probably had similar atmospherics and wind patterns, today, the very top layer of Venus' clouds zip around the entire planet every four Earth days, propelled by hurricane-force winds traveling well over 200 miles per hour. The image below shows the hurricane like winds of the upper atmosphere.

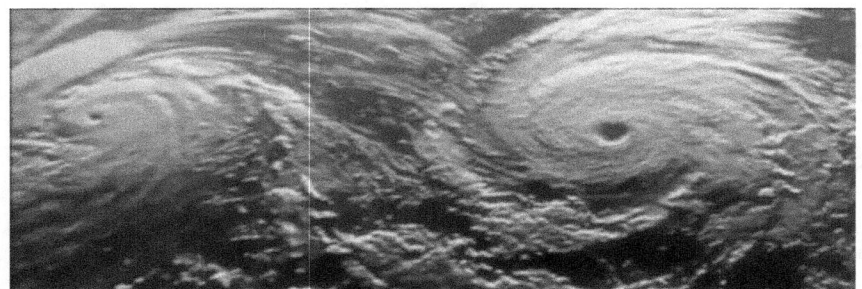

This super-rotation of the planet's atmosphere is 60 times faster than Venus itself rotates and nobody knows why it happens. Down at the surface, winds are only a few miles an hour, but if the destruction of Venus had been millions of years ago, we would find no signs of life at all.

While we must believe, a Venusian day and an Earth day once were similar, Venus now takes 243 days to turn once on its axis, and it takes almost 225 days to travel once around the Sun in orbit. While it is weird, a day on Venus is longer than its year. The chart shows the size to rotation differences of the planets. Please notice Venus is completely different showing something happened to it.

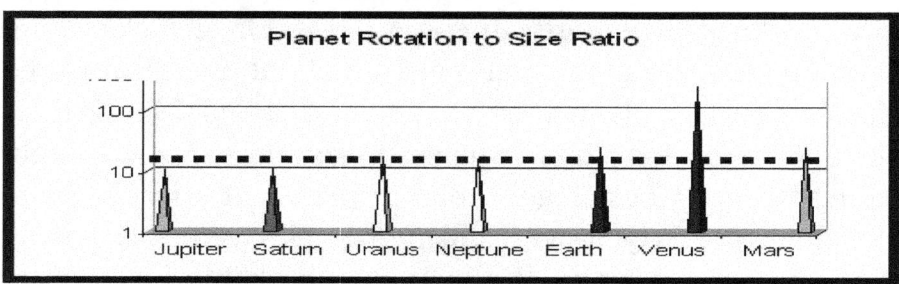

If that's not strange enough, the rotation of Venus is backwards. Seen from above, all of the planets in the Solar System rotate counter-clockwise. But that rotation on Venus is backwards, so it's going clockwise. On the surface, if you could stand the pressure and temperature, you would see the Sun rise in the west and then take over 100 days to travel across the sky and then set in the east. So why is the rotation of Venus backwards? The atmosphere is caustic, the temperature melts metals, the atmospheric pressures are mind boggling so who in the world would ever claim Venus had people living there?

What Happened?

Beautiful Place- At one time, Venus rotated about the same as Earth was slightly closer to Earth and was filled with all types of plant-life. Massive oceans and waterways held a vast array of animal life. One of the plants native to Venus we believe is the one we call Venus Flytrap. I'll explain later. This is not fiction as we still find the remains of all of this today as the Destruction was not long ago. We will look as some of the features that are similar to those found on earth so you will have a chance to make up your own mind.

Shattering Moon--One possibility is that Venus used to have its own moon similar but much smaller than Lunar. The Electrical potentials of the two planets got so different, that one day the Venusian moon came between the 2 planets and it shattered. When the pieces struck Venus, almost split in half and the impact shifted the rotation, slowed it down to almost nothing, and the residual electrical difference pushed Venus slightly closer to the Sun. We will look as some of the

Venus Gets a Tail-The electrical plasma was so dense, it became visible just like a Florescent Light and people on Earth saw Venus as the planet with a fiery tall for years. Around the globe stories about the planet having a tail showed it could been seen all over the world for a time. Even today the SOHO satellite registered the remains of the Plasma threads coming to within miles of the Earth's surface whenever the 2 planets come within about 5 million miles.

Meteorite Destruction-A large number of the pieces of the Venusian moon and the Venusian planet itself headed for Earth and struck with a vengeance 11 thousand years ago causing ½ million craters along what had been the Earth's Equator and is now the East coastline of the United States. The quantity of Meteors caused massive forest fires and entire civilizations may have been ended by the planetary assault.

I hope, by now, you are not thinking, this is all made up and now that the entire planet is, essential melting, there is no way to have any evidence. While that would be true if Venus caught on fire a million years ago, the relatively recent destruction allows us to see what Venus used to look like.

Venus Crack Anomaly

Just one more obvious thing needs to be discussed with respect to this horrible event that occurred some 11 thousand years ago. The reason we know when Venus was almost split is that whatever happened on that planet greatly affected our own. Thousands of meteoric bombs hit our earth at the same time Up until that time there was a tight union between our 2 planets and there is still are plasma streamers visible today showing the once close association. Venus almost split apart just like Mars did 400 thousand years earlier. When we look at the topographical map of Venus we find very deep cracks almost half way around the planet. It looks like Venus was almost split in two from something pretty catastrophic like its moon hitting the surface. For those finding it difficult to make out the gash, I have highlighted the extremely deep and long fissures that go almost half way around the planet. If the incident had been even the slightest bit more intense, Venus would be a much smaller planet today and more asteroids would be whirling around. The first image shows the Venuisian moon possibly hitting the surface and millions of pieces of debris spreading out towards the Earth and beyond.

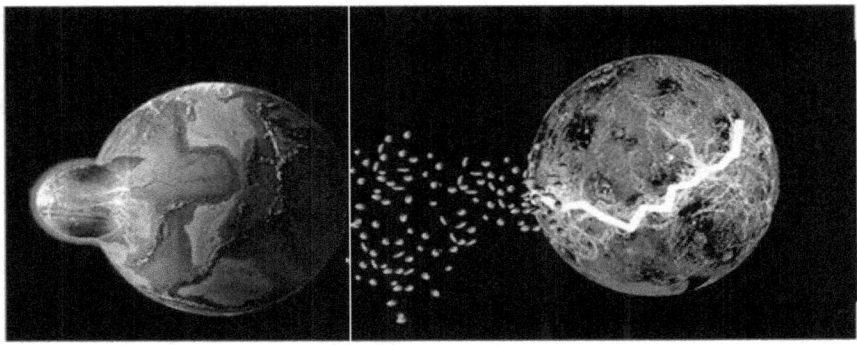

Another photograph, next, may even show the gash better. It's weird that you were never introduced to this obvious feature before, isn't it? Even the Bible tells of its doom. The Bible calls the planet "Rahab" and God destroyed Rahab because of its vanity and because this bad guy named Satan had colonists on the planet. On earth, the massive bombardments caused the earth to shift which spelled doom for many of the animals of that time. Extinction here was nothing like the massive destruction on Venus as its entire ecosystem was destroyed and it essentially ignited, melting just about everything including people.

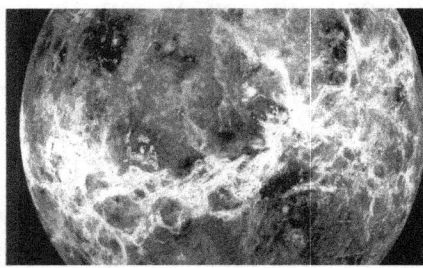 

The Magellan space probe photographed 98% of the surface of Venus. All the large craters, shown as light circles, were found to be along the equator of the planet. [See image above right] What that means is something moving around Venus along its equator exploded. The most reasonable choice is the planet once had a "moon" revolving around it's equator and it blew up not too long ago as a massive electrical discharge established a massive plasma ribbon between Venus and Earth.

Blazing Tail Anomaly

I told you previously that there was an electrical difference between Venus and Earth. It got higher and higher and then catastrophe struck. Just like turning on a florescent light, the vibrating electrical difference cause a current to flow as a plasma and it illuminated, but a Venusian moon or similar object got in the way and shattered. Plasma trails can be and are produced from violent electrical disturbances occurring on planets, if they are significant enough to cause quick atomic ionization. This atomic ionization would be expected if parts of the planet were yanked away or hit with a huge electric field or hit by a huge lightning bolt. Additionally, we must understand that plasmas, although they are basically lumps of gas, do not behave like gases. They develop structure. When a huge variance in electric potential gets too high it produces a huge electric current that, in turn, causes the ionization. Sorry for the details, but I'm telling you for a reason. The current flow also causes these huge magnetic fields that, finally makes the something we call plasma filaments [tails] that twist together into things that look like "Blazing, wavy, gas ropes". As long as the current continues, the structure of the plasma remains intact. Sometimes these ropes become very visible. If Earth were affecting Venus, there would be plasma tails between Earth and Venus. In Kohistan, there is a cave full of ancient cave drawings. Researchers insist that a planet and star pattern depicted shows the alignments of stars as they were about 11 thousand years ago. Here is the thing that I particularly like--- there are lines drawn between Venus and Earth. They look like Plasma tails between the two planets. While this is evidence that these plasma things were around about 11 thousand years ago, we have found that the plasma tails were not visible only in these ancient times. The event essentially destroyed Venus, but it plasma tail as a remnant of the discharge remained and still can be sensed today.

During the first 5 thousand years after the destruction, the visible tail was easily distinguishable as we can determine by ancient writings. By erasing Venus's tail in our history books, quasi-historians could weave an alternate tale about human history that is so very wrong. Here are just a few of the descriptions of Venus. Try to pull out indications of a fiery tail. It will be very difficult so one can understand why this important factor has been erased.

European Venusian Tail-The European countries remembered the devastating event and wrote about it. Greek stories are filled with details. Here is one. According to Greek legend, *"A blazing star almost destroyed the world with fire before it became Venus."* [Although it is difficult to interpret, I believe they are talking about Venus turning into a blazing star.]

South American Venusian Tail-The People of South America remembered and wrote about it. The Inca legends tell the story. The Inca called Venus the *"Wavy haired planet"*; [This also seems difficult to interpret. Could wavy hair be flames shooting from its surface during a time when the Inca were around?]

Aztec Venusian Tail and killing meteors-The People of Central America remembered and wrote about it. This is from one of the Aztec legends. The Aztecs called Venus *"The Star that smoked"* and said, *"It once passed by the world blazing and killing many people."*

In the Mayan Dresden Codex, the god of Venus is depicted with shooting darts. It seems to me that if something is shot away from a planet, it would have been meteor-like. The picture below is from one of the Dresden pages. You guessed it Venus is in the middle.

Blackfoot Indian Venus tail-The People of North America remembered and wrote about it. Let's see what the Blackfoot had to say. According to their traditions, *"The morning star [Venus] put on a scarlet cloak [sounds like it turned red.] and appeared before a woman on Earth that he loved. She went into the sky with him, but was warned never to look back. She did, of course, and was ordered to return to Earth."* [The return was a mess if we believe the other histories.]

Ute Indian Venus Meteors-The Ute Indians tell us the same thing in their verbal history. *"The sun was slivered into a thousand fragments, which fell to Earth causing a general fire. Then Ta-wats fled before the destruction he had wrought. All were consumed; until at last, swollen with heat,* *the eyes of the god burst and tears gushed forth in a flood which spread over the Earth and extinguished the fire."* [This flood appears to be talking about the worldwide flood at the end of the Pleistocene so we can date this description. As far as the sun bursting, I personally believe it was Venus and not the sun.]

Egyptian Venusian Destruction-In Egypt, the event was known and written about. Sonchie, the high priest, told Solon, a Greek historian, about events before the flood. He wrote, *"Many are the destructions of mankind that have been and shall be. The greatest are by fire and water. During long intervals, there are <u>deviations of the bodies that move around the Earth in the heavens and the consequence is widespread destruction by fire</u> of things on the Earth."* [The fires must have been everywhere when the Venusian moon split apart. The comment that it was one of the "Normal bodies that moved around in the Earth sky" limits the body to one of the close planets. Of course, the closest is Venus.]

Sumerian Venus tail and destruction-The Sumerians made record of the blazing tail of Venus. Their goddess named Inanna was associated with Venus and the information is the same as recorded by all the rest. *"To the queen of the heavens Inanna [Venus], to her who <u>filled the sky with her pure blaze</u>. The <u>luminations are as bright as the sun</u>. Who initiated the flood-storm? You roared in the heavens and Earth. You smote the flesh of the people."* [The blaze of Venus filled the sky, roared across Earth and smote the people. I think the only way Venus could smote the people is if its moon exploded and pieces fell to earth as a huge meteorite storm.] *"She [Inanna/Venus] who causes the heavens to rumble. She who shakes the Earthquake. She cried toward heaven and Earth, My hair will whirl in heaven for you. You flash like* *lightning over the highlands. You throw firebrands across the Earth. You split apart the mountains."* [The hair extending sounds like a reference to a comet tail or a blasted away section of Venus that hit the Earth. Firebrands hitting the earth sounds like meteors to me.]

Assyrian Venus tail and destruction-Assyrian literature tells the same story. This time the goddess is named Ishtar, but it is the same. *"To the pure flame that fills the heaven, who shines like the sun 'Ishtar" [Venus]—I ran battle down like flames in the fighting. I make heaven and Earth shake. I trample the Earth. I destroy what remains of the inhabited world".* [To destroy the remains of the inhabited world, there must have been something substantial that happened with Venus.]

Arabian Venusian Meteors -Coptic texts date the event for us in the Age of Leo. The ancient Arabic text called "Bundahishn" tells us the following: The Ancient Coptic text said *"There would be a great fire and flood coming out of the constellation of Leo".* [This not only describes the event but places it in the "Age of Leo", 11 to 13 thousand years ago.] It goes farther indicating that the beginning of world history was around 11 thousand years ago and some of the major deities were born during this event. The beginning of history must have meant that there was a destruction period just before that time.

Phoenician Venus tail and destruction-Phoenician texts describe the event, but this time the goddess is Astarte, the Phoenician version of Ishtar. *"See, Astarte" [Venus], she descends into a pool as a fiery falling star".* [A beautiful description for a terrible disaster.]

Persian Venus caused fire and flood- Mandaean Texts from Persia give us the same information. *"150 thousand years after man was created, the whole Earth broke out into flames and only 2 escaped."* It continues by saying *"They had children and, of those ancestors, Noh [almost like Noah] was the one that survived the Flood."* [The Earth being filled with flames could have been from the huge quantity of meteors from the explosion, but clearly this event occurred well before Noh survived a worldwide flood.]

Indian Venus Tail and destruction-The Indian writers also informed us of this terrible calamity. The people remembered *"Venus sweeping away the stars."* Indian literature states the following, *"Her [Venus's] anger grew so terrible that she transformed herself, grew smaller and black. On a blind rampage, <u>she was killing everything</u> and everyone in sight. Her <u>hair is wild, her eyes red.</u> The world trembles and cracks under her tread. <u>Her dark hair flies in the sky sweeping away the sun and stars.</u>"* [Again, we read about the comet-like tail and so many meteors that the sky is darkened. Also notice Venus got smaller as the planet went closer to the sun.]

Chinese Venusian Meteor Storm-The Far East writers also informed us of this terrible calamity. The people remembered Venus sending down a huge meteor shower. The Chinese writers said the same thing, *"There was a time when a planet [Venus] <u>approached close to the Earth, causing great showers of stones.</u>"* Not too many of the planets could have come close to earth. The moon of Venus is my guess. Venus was depicted as a dangerous *"fire spitting planet"* according to another Chinese legend.

Pacific Island Venusian Meteors- Even the people of the Pacific remembered Venus sending down a huge meteor shower. Venus was depicted as a dangerous, *"fire spitting, planet"* by the Samoans. It is like reading the Chinese version. What would have given them that idea?

Meteorite Physical Evidence-Not only did people write about the event, but also, there is a huge amount of physical evidence in the form of thousands of craters left when the flaming masses hit the Earth. The pieces hit places around the world. We know when they hit, we know that the explosion that caused them was reasonably close, and we know that there were many thousands of meteoric chunks that hit the earth at the same time. Large amounts of "meteoritic mass" and an estimated 500 thousand strange indentions, strongly believed to be from massive meteorite showers have been found around the world that date to the end of the Pleistocene era, about 11 thousand years ago. Large quantities have been found in Alaska, Siberia, Bolivia, and Netherlands. Guess what! The time period for the destruction of the Venusian moon is about 11 thousand years ago. If they both happened about the same time, there is a good possibility that they were the same event.

Glass Evidence-Tektites are small pieces of glass formed as a meteor strikes the ground and melts the surrounding area. Many have been found in sort of an "S" shape and distributed over large portions of the Earth. Some were found embedded in fossilized wood, in Australia, others were found in Vietnam and still others were found in the Indian Ocean. Several dating methods were used including Stratographic, Carbon 14, and others. They showed that most of these pieces were deposited around 10 thousand years ago. Ok! Maybe the ones inside the fossilized wood came from an earlier strike, but most were Pleistocene Era events just like the Venus moon blast.

New Zealand Evidence-Today, huge quantities of metallic meteorites as well as objects called "china stones" can be found everywhere on the island. Inside the stones are the remains of burned up Pleistocene type material, which dates the event to between 10 and 20 thousand years ago. [I suppose you think these came from the Venus moon strike just like me.]

So, we read over and over again, Venus had a flaming tail and not long after that started, thousands and thousands of rocks descended on the earth and many were killing in the resulting fires that swept across the lands. There was a consensus that the meteors actually caused the Earth to become unstable and this caused the flooding of the entire world. Certainly, we should be able to find some of the craters and we should be able to date them to the time of the Venusian Destruction.

That brings us to the next anomaly ignored by historians this is a collection of craters called the Carolina Bays.

Carolina Bays Anomaly

The east coast of the United States was pelted with many objects. There are still an estimated 500 thousand meteorite indentions called "Carolina Bays", which mark this incredible event in history. One hundred and forty thousand of these holes have diameters of over 500 feet. Just think about how afraid the people of that time were as they essentially saw the sky fall all around them. The picture below shows the major areas where these objects have been found in the United States. On the other side of the world in Australia we find more of these same holes.

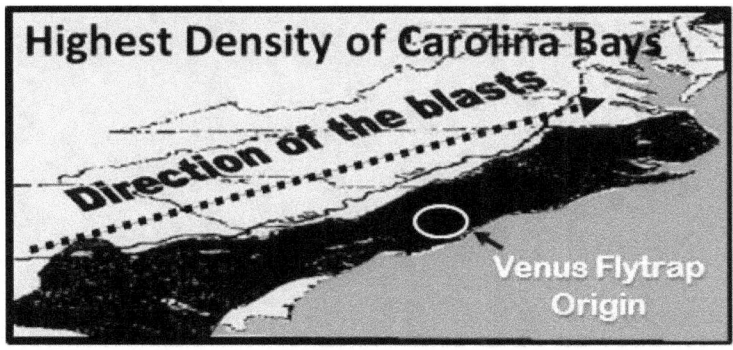

These generally date around the same time [end of the Pleistocene]. The evidence shows that the Venusian moon met its end at the same time that these 500 thousand holes appeared. Some of these indentions are very large and have diameters that are thousands of feet across. So, it wasn't just a little meteorite storm. The direction of the blast will be important later, so I put it on the drawing to get you used to seeing its direction which is about 30 degrees off axis from our Equator [as if the Earth equator had been different 11 thousand years ago].

The Carolina bay incident was a huge onslaught of meteors striking the Earth, which caused holes everywhere. Don't just take my word on this. The collage below shows a quantity of these things in a small area. These holes are found along almost the entire Eastern coast of the United States. *Isaiah 14:12- How art thou fallen from heaven, O **Venus** [the morning star]! How art thou cut down to the ground, which didst weaken the nations!* [As ½ million meteors set vast areas of the Earth on fire.] The images show a tiny fraction of the ½ million craters and the locations of the largest ones.

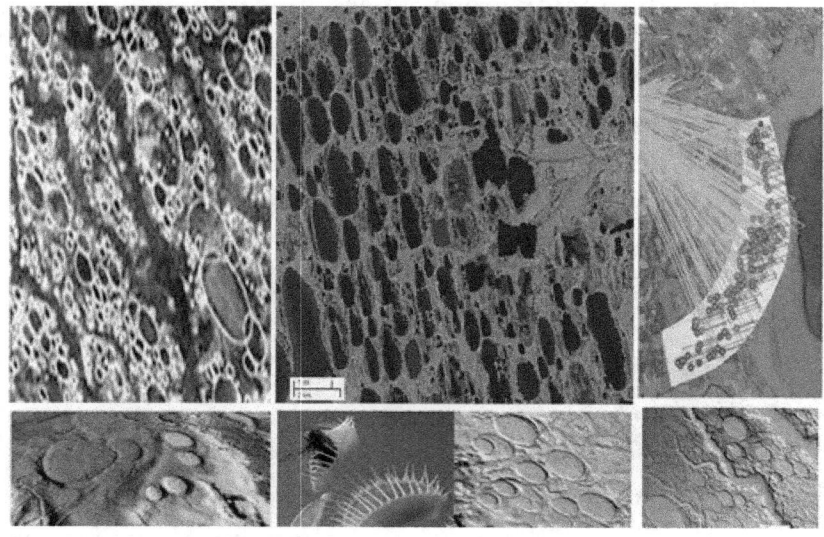

Australian Carolina Bays-If we look on the other side of the world we get more confirmation as thousands of round saline lakes span the southwestern part of Western Australia, the image below left shows how an equator along the eastern coastline of USA also would travel around to the south west portion of Australia. To the right is a satellite image of some of the crater lakes.

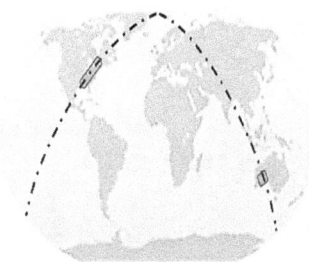 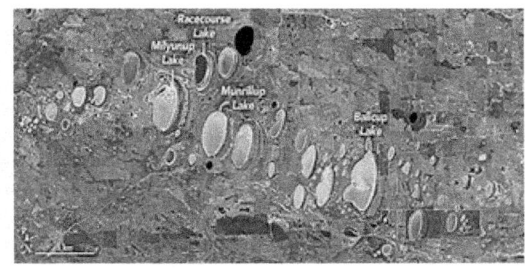

While all this was going on, please don't go into how you were told all the humans living in the Pleistocene lived in caves, dragged women around by the hair, and said "ugh" all day. From dozens of ancient texts, radioactive evidence, massive human mutation evidence, and evidence of how fast civilization expanded after the Pleistocene Extinction we must believe people around the world were highly civilized. The book Jasher indicated 1/3 of the population of the world died BEFORE the Pleistocene extinction and worldwide flood of Noah. Possibly the massive fires from these meteors caused some of the death.

Venus Flytrap Anomaly

Possibly you noticed the image of a Venus Flytrap at the beginning of this section. If you saw it, I'm going to tell you something very strange. Even if you didn't notice it, I'll tell you as this book tries to be an equal information book. What you probably didn't know is that this carnivorous plant is indigenous to the Carolina and nowhere else. The Carolinas are the middle of this massive meteorite grouping. The fires that occurred during the Meteor storm would not have provided a good timing for a new species of plant, but some of the spores could have been captured in the meteors. Therefore, what we are seeing might well be one of the indigenous plants of Venus during the Pleistocene Age.

Travel Anomaly

When I'm talking about travel here, I'm really talking about space travel. I'm not spending a lot of time on this subject as I have an entire book of flying anomalies but let review. We had Biblical testimony that there were, possibly, colonists living on Venus before it was destroyed. The way to get people up to Venus is to use a flying machine of some kind.

"Ramayana"- The "Ramayana" said the following, *"It was a self-sustaining flying city that* <u>*traveled in outer space*</u>*" -- "One of these cities was named Hiranyapura (city of gold)"*

Amsu Bodhini"- Information about planets was found in the book "Amsu Bodhini" which was not decoded until 1931. It contained information about *different kinds of light, heat, color, electromagnetic fields, solar energy, capability to send messages by cable, and* **machines to carry people to other planets.**

On and on we could go as flying machines of all type we extensively used for transport, war, and space exploration before the Bharata War messed up our brains.

Bible Describes the Pleistocene War

Jubilees 5:1-9-*And against their [ancient people] sons went forth a command that they should be smitten with the sword-- --And he sent his sword into their midst that each should slay his neighbor, and they all began to slay each other till they fell by the sword and were destroyed from the Earth. And their fathers [ancient people] were witnesses of their destruction, and after this they were bound in the depths of the Earth.*

Jubilees 5 and 7 -We need to go no farther than the Jewish book of "Jubilees" to investigate the amount of fighting during this time. According to this account, essentially ALL men and giants died. *And lawlessness increased on the earth and all flesh corrupted its way, alike men and cattle and beasts and birds and everything that walks on the earth -all of them corrupted their ways and their orders, -And they began to slay each other till they all fell by the sword and were destroyed from the earth. --And <u>He destroyed all from their places</u>* [Including Venusian outposts], *and there was not left one of them. ---and the Giants slew the Ancient humans, and the Ancient humans slew the Eljo [other half breeds], and the Eljo mankind, and one man another. -- And He said 'My spirit shall not always abide on man; for they also are flesh and their days shall be one hundred and twenty years'. And they began to slay each other till they all fell by the sword. The Nephadim slew the Ancient humans, and the*

Ancient humans slew the Eljo, and the Eljo slew mankind and one man slew another man. [All the various humans were involved in these terrible wars according to the ancient texts. This might be the best description of how the Civil War escalated into a true World War. We will see that the war also spread beyond the earth.]

Enoch 10:13-14-*Destroy the children of fornication, the offspring of the ancient people, from among men; bring them forth, and excite them one against another. Let them perish by mutual slaughter.*

Generation of Adam 11:3- *"Leboa, Daughter of Tamar, devised a "Sword of Light" which penetrated the wall of defense around the city of Haner and began to drain the power from the wall."* [Some type of Laser beam apparently penetrated whatever the wall was and drained its "power". Sounds almost like the science fiction of today, but these preflood, world war weapons must have been amazing.]

Hesiod [Greek]--After the Greek description of the heaven wars, Hesiod continued with the description of wars on the earth between the ancient people called "gods" and just about everyone else. *All the gods were divided in strife, even to mingle storm and tempest and already hastening to make an <u>utter end of the race of mortal men</u>, declaring that he would destroy the lives of the demi-gods, that the children of the gods should not mate with wretched mortals.*

Burst of Radiation Anomaly

We know that many animals and plants became extinct at this time. Then an indicator tells us more. *Uranium concentrations in coral jump by almost 300%.* Also, we find marked increases in **nanodiamonds,** magnetic spherules [tiny balls], and carbon spherules at the end of the War with a major increase in charcoal around the middle showing fire and general war conditions. The nanodiamonds indicate substantial heat like nuclear explosions. The darker area below represents the Dryas [10 to 11 thousand years ago].

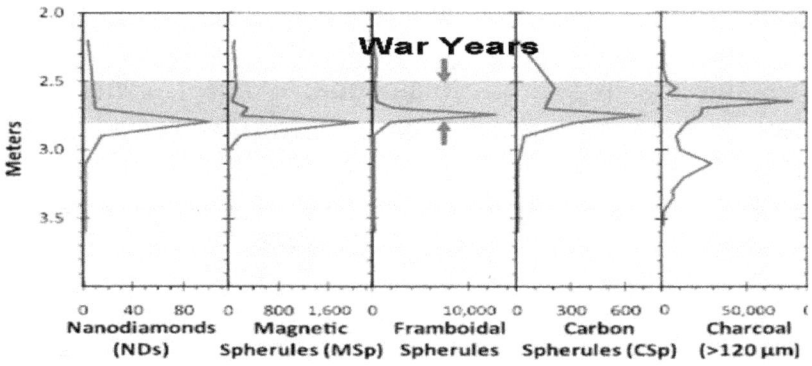

Hopefully, you are at least intrigued by the data that teachers have not provided to students so that they could make some logical determination of the events that molded Venus, sunk the island of Atlantis, elevated the radiation level of the earth, made the temperature drop suddenly, caused massive amounts of mutations in humans during a very short period of time, cause many dinosaur bones to be radioactive, caused the plasma strings being seen by Soho, caused 500 thousands craters along the East coast of the United States, and was described over and over again by people around the world. Please do not just believe CONSENSUS. The scientist using this horrible tactic have an agenda and it is not to inform people about the truth.

Consensus Dryas

In this section, I'm going to tell you about something consensus scientists call the Young Dryas. This was a strange time just before the end of the Pleistocene Extinction and worldwide flood. Strangely, the war, massive radiation levels, flying and Venus destruction all seems to fit together, but consensus scientists want to ignore the conditions as simply say "Dryas are a mystery". The graph following shows a dramatic in temperature between about 11 thousand and 10 thousand years ago that quasi-scientists simply call Dryas. While we have been timing the 10-thousand-year event as the end of the Pleistocene; what in the world happened a mere thousand years before the Pleistocene Extinction?

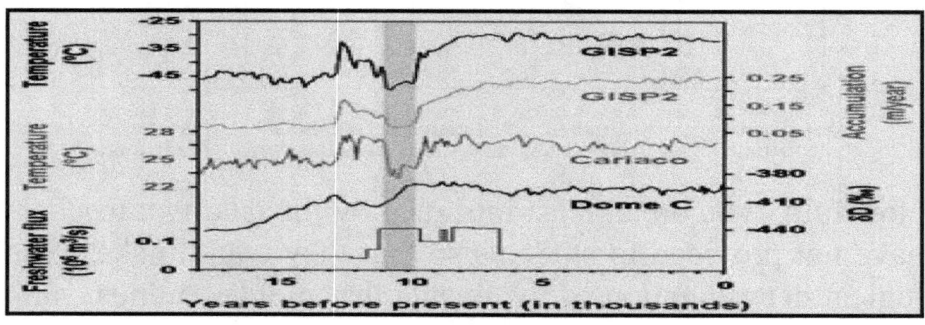

The top two ice core samples in the preceding image are from Greenland and the last 2 are from Antarctica to show this was a worldwide event. While we don't know for sure exactly what happened there are indicators that strongly suggest something terrible happened on both planets not too long ago. Certainly, it was well after the time that Mars sucked out the Pacific Ocean and the timing of the Carolina Bays tells us it was just before the end of the Pleistocene. Here are the things we know that are starting to make sense or at least I hope they are.

Craters- ½ Million meteors came from somewhere around 10 or 12 thousand years ago. --- remains of an exploded moon---

Temperature-There was a massive drop in temperature around the world about 11 thousand years ago. ----- No telling what happens when ½ million meteors hit and set an entire continent on fire---.

Radioactivity- A sharp short-term rise in radioactivity about 11 thousand years ago. --- possibility of Nuclear war----

More Radioactivity-We also are finding unfossilized dinosaur remains that are highly radioactive and knowledge of 16 nuclear processing areas in Africa that predate the end of the Pleistocene. --- This sounds like nuclear war-----

History-There are many depictions of a massive worldwide war just before the worldwide flood. This includes stories from our Bible indicating as many as 1/3 of the population of the world was killed. ---- Nuclear war could kill 1/3 of the population if you add ½ meteors----

Nephadim War- Some of the depictions of a Pleistocene War include a race of Giants called the Nephadim who were all killed along with all of their descendants. This was timed to be during the time of Lamech who ruled around 15 thousand years ago by some <u>histories</u>. --- Timing seems right and Giants might be good fighters-----

Plato- tells us of a group of Islands that sank, before the end of the Pleistocene. He timed the submersion to be about 12 thousand years ago. --- We can believe ½ Million meteors could have caused a massive change in weather an even began the slow change of the Earth's axis which could make Island civilizations submerged forever. ---

Venus- We are now finding many signs of war on the Planet Venus. These are being timed as "recent" by astronomers. --- This matches the 11-thousand-year timing---

Venus Rotation- In fairly recent times, Venus rotation shifted 90 degrees from all the other planets. ----One thing that could do that is being hit by you own massive moon at just the right angle to flip the planet sideways. -----

Venus Craters- We are finding that most of the major craters on the planet fall along the equatorial region. ---This is as if whatever caused them at one time was a moon. ----

Planetary Colonization-Indications that the planet Rahab [Venus] was colonized and part of a war are found in several ancient documents. ----- Physical evidence confirms them, ---

Destruction of Rahab- Clear indication of the destruction of Rahab can be found in our Bible and other histories.

Weaponry-Descriptions of the use of mighty weapons in a Pleistocene War are found in a number of Bible related works.

DNA- We are told by Haplotype scientists that almost ½ of all major DNA mutation occurred 10 to 12 thousand years ago and most of the other mutations occurred around the time of the Bharata War. --- Nuclear War-----

Plausible Explanation

These things seem to tell us there were civilized people on the earth during the Pleistocene who experimented with DNA remaking dinosaurs that had once been extinct from the ending of the Cretaceous. Desire for power abounded and there was a war before the worldwide flood and many were killed. At some time, nuclear weapons were used. During this time, the planet Venus was livable and colonized, but some terrible even destroyed the planet and sent massive amounts of meteoric debris to the Earth. This meteor shower upset the weather patterned and Islands became submerged. I know some of this is not being taught in our schools and I don't have a problem with that so much, but what I have an issue with is that the things we know are not being taught.

Kingdom after kingdom became accustom to war. Some flying ships were sent to remote sites on the Moon and Venus according to many sources. Venus was not like it is today. It had green field, huge rivers, massive oceans and air. According to the Biblical account the Planet Venus was called Rahab or "Vain Place". It also states that one of the warring generals of this time originally named Gadrael was massing a large group to assault "heaven". The book of Jasher indicates that "*1/3 of the population of the entire world was lost in the World War events, but that was not the end of it*".

The ancient historical works tell us that the planet Venus was much larger in the sky than it is today and both planets affected one another somewhat like the earlier events of Mars, but not to that extreme. Venus would have been the best colonization port as it most likely had sufficient atmosphere, oxygen and water to support a large civilization. The remains of huge river systems, rolling hills and valleys can still be seen today, but there are no more people and there is no more breathable atmosphere, in fact, it is almost all CO_2. There also has been a huge temperature jump to over 800 degrees, the air pressure has jumped to 90 times that of the earth's, and thick clouds of Sulfuric Acid now cover the surface. You can certainly believe that anything that had been alive on the planet was destroyed in the transition. Another thing that is noticeable is that the rotational speed slowed to almost nothing. The logarithmic chart below shows the rotational ratios to planet size of the other planets. Notice that Venus is way out of place. It has almost no rotation. In fact, Venus is currently rotating about 1/100th as fast as its sister planet Earth.

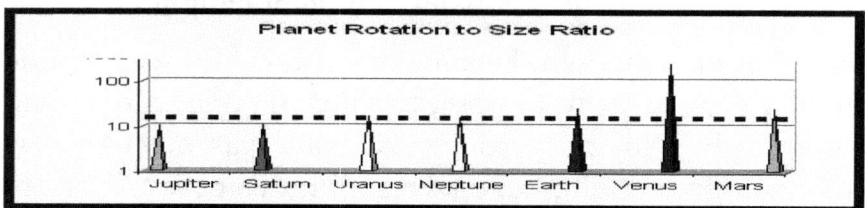

Something catastrophic happened on Venus but we will only investigate it superficially, as it affected those on the Earth during the World War before the Pleistocene Extinction. The image below left shows what Venus looks like today with most of the massive craters along its equator?

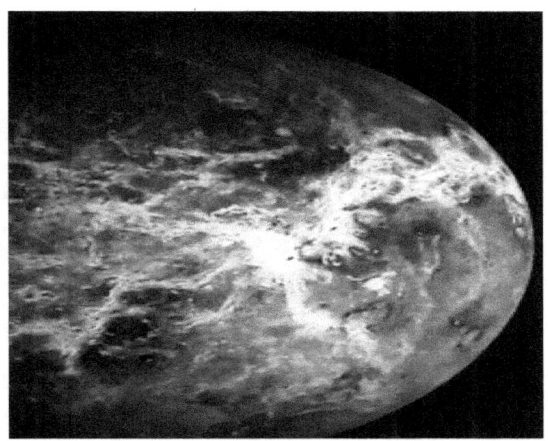

The moon hit with such force the planet was almost spit in two. We may even know where the impact as X marks the spot [Preceding Image]. This drove the rotation to an almost stopped condition, causing the atmosphere to build until the pressure level hit unbelievable levels, structures melted, and rivers died up immediately leaving well defined images of their once great streams.

Why- Let me explain a little more about Plasma. The Earth "potential" can be millions of volts different than a nearby object. This happens naturally as we spin around the sun. There is no place to discharge it so no one notices. If Venus had a slightly elliptical orbit or somewhat closer to Earth before destruction hit we have a problem. Whenever planets come close to one another, the massive difference in electric fields of the two planets could, quite easily, establish this Plasma stuff, which is sort of like and electrical discharge path without a wire. We use these things all the time in Xenon strobes and Fluorescent light bulbs, but when we are talking about an entire planet discharging, it is pretty nasty. One reason to talk about plasmas is that the Soho Satellite registered a substantial indication of a plasma "electrical discharge string" emanating from Venus and going towards our planet. The image below shows the satellite position when it captured this unbelievable emanation.

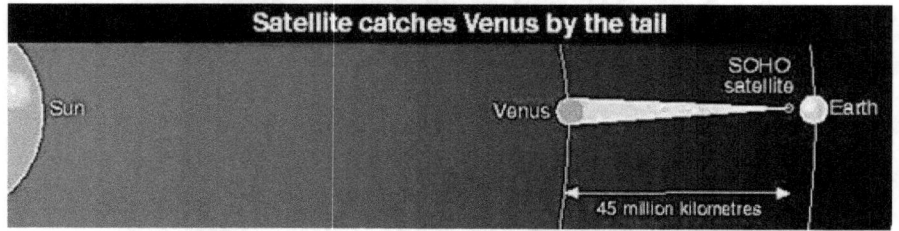

The image shows the information from the SOHO satellite. Venus is outputting the remains of an electrical plasma ribbon that once connected both our planets. Venus had its moon get in the way and it was too much for the tiny object.

Plasma-If the energy was discharged to a planet it might be bad, but if a moon got in the way, the entire structure could be blasted into thousands and thousands of meteors. If the meteors struck Venus at an angle opposite the rotation, it would slow down as we see today. The end of a livable Venus may have happened in a matter of minutes.

Lava and Mountains

By far the most noticeable thing on the planet is lava. The collage shows just a few of the volcanos that pot mark the surface. While they appear to be spewing lava out for hundreds of miles making it hard to walk even with shoes on, this odd thing is the eruptions happened all at one time and the outcome is frozen in time. The detail left by the eruptions show that the incident did not happen long ago. Scientists tell us these are new eruptions.

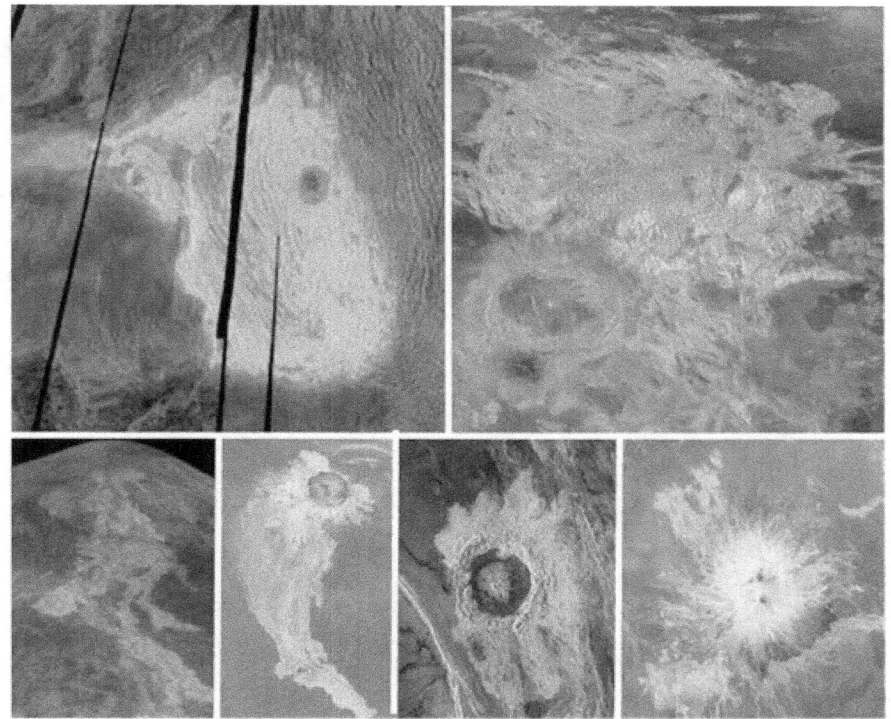

Besides the Sulfur Peppered Carbon Dioxide air, we find what looks like snowcapped mountains. What appear to be snow-capped mountains is actually metal topped mountains where the metals have melted to build a highly reflective surface feature. One of the many beautiful mountains is shown next.

So, what? Lava and mountains don't help inhabitants. Certainly, the fact that the lava flows are all recent occurrences shows that there was plenty of time before the massive lava flows that would have allowed people to live on Venus, what about water? Scientists tell use there are traces of water on the planet today, but that would not sustain a community.

Venus River Anomaly

Some may tell you Venus has no water, but at one-time Earth and Venus were very similar.

Something happened on Venus and it wasn't millions of years ago as you have been told. It happened about 11 thousand years ago by all evidence. Before then it was inhabitable and most likely, inhabited by humans. Venus holds a special place in earth's history. The evidence suggests that the landscape of Venus had many winding rivers and a lush atmosphere.

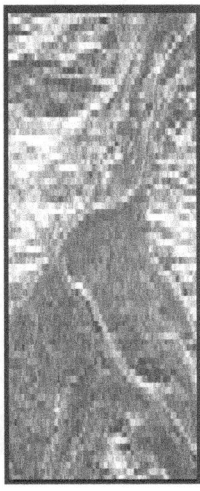

One of the winding rivers is shown to the right above. Two events occurred to change all that and NO it wasn't the "GREENHOUSE effect that we were told in school.

Scientists know that something was orbiting Venus along its equator and it exploded. The debris pelted the planet surface along the equatorial region and huge slivers of the exploded mass bombarded the earth as well---again, along the equatorial path. Besides this horrible destruction of the Venusian Moon, there are also significant signs of controlled blasts on the Venusian surface like those seen in aerial bomb raids. This suggests that Venus was caught up in an ancient war with people living on the earth. First, we will look at what happened before it became an inferno and later we will discuss the final blow. The planet was, most likely, beautiful and green before the wars hit.

The picture below shows one of the many, now empty, river deltas. Notice how the delta is still perfectly formed showing that the event that changed the environment did not happen a long time ago. Now the water has all gone, but the riverbed is still visible along with the tell-tell signs of war. Below is still another dried-up river. Along its shore are the unmistakable signs of civilization with rectangular areas that, most likely, had once been buildings and regularly spaced elements that do not look natural.

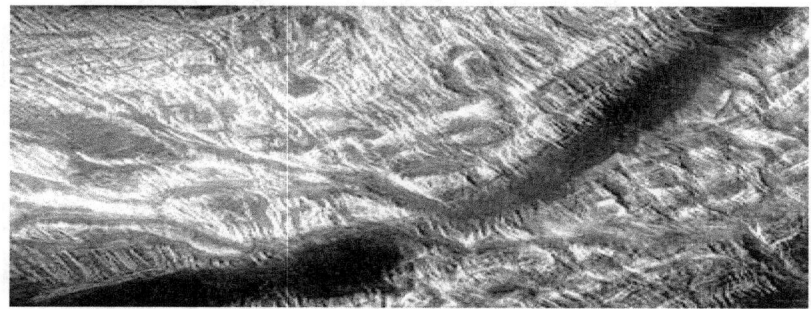

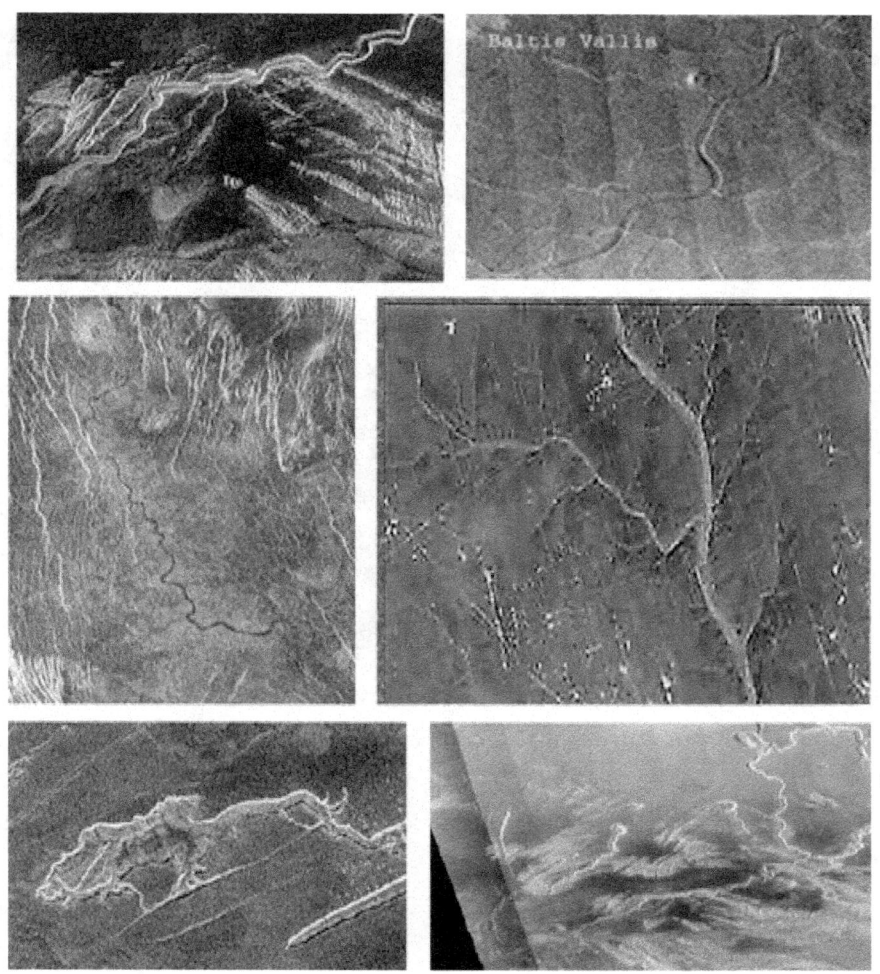

The next thing we find is what appears to be remains of buildings beside massive waterway now completely dry. The next image shows the anomaly. I enhanced the areas of interest to show how it used to look.

Hopefully you are beginning to see Venus as a beautiful planet that was quickly destroyed a few thousand years ago. If you have water, and building next to rivers, could there be more that are not completely melted by the tremendous pressures?

Buildings

Just because the planet used to have rivers doesn't mean people lived there, but it seems that every few months someone finds another geometrical shaped, building or what appears to be buildings that once must have peppered the landscape before the destruction. Certainly, these are all melted and ruined today, but may you can make out the previous grandeur. The high walled U-Shaped building is followed by an "H" shaped structure and 2 square buildings. The last on the top row certainly is not a natural structure.

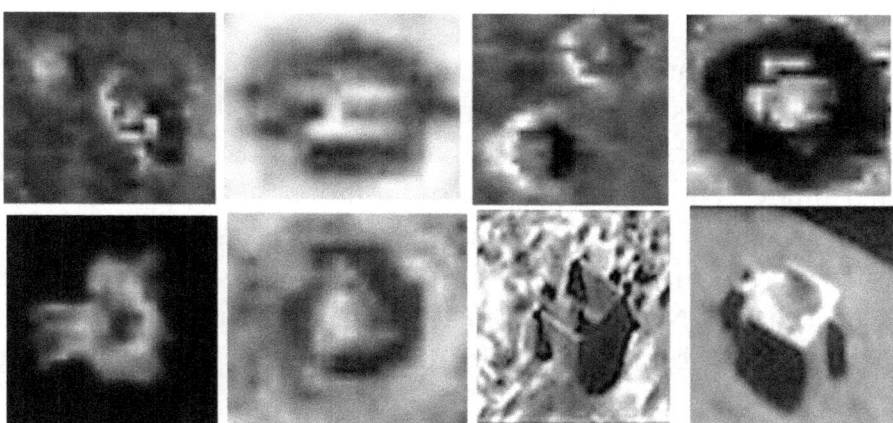

The second row above has another geometric shaped structure followed by a triangular building and 2 square high walled structures. The top row of the following group shows a geometric complex, massive high walls and then two square buildings that are adjoined. The 4th image shows walls and a square structure and the last on the top row looks like it has a tower of some kind.

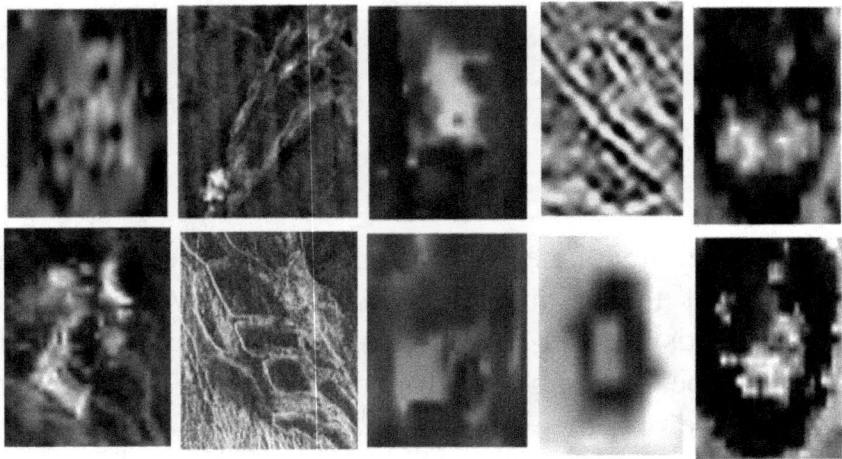

The second row above starts with a multi-component grouping of buildings followed by what could be a three-stepped set of square buildings. The three more structures with squared sides end this group.

Please understand the detail is substantially less clear through the thick atmosphere than images taken of the Martian and lunar surfaces but we can still see suspicious items that look man-made. The following are even more that appear to have been used by colonists at one time. The first appears to be some type of high walled building while the second structure is triangular. Notice the third image on the top row has perpendicular building made in the shape of a H just like a preceding one that I showed.

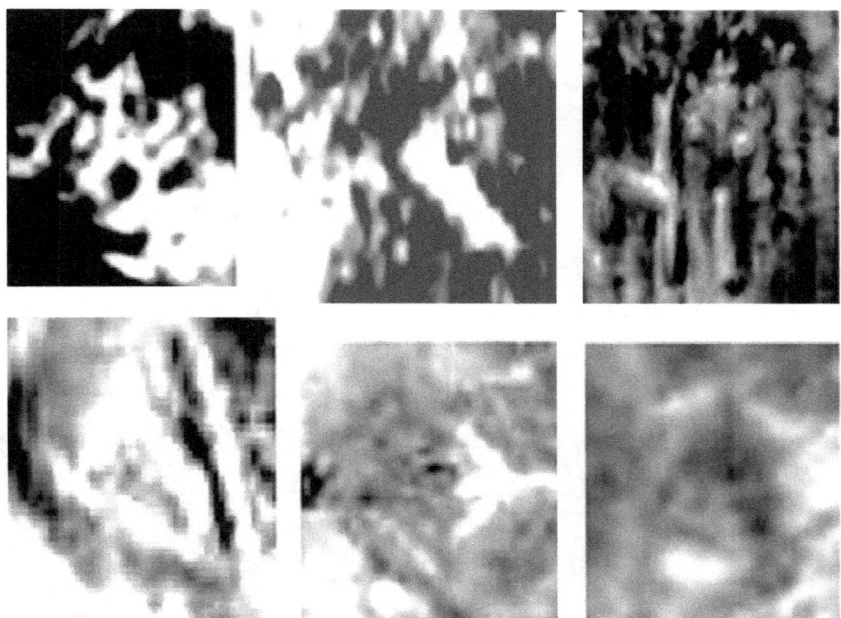

The bottom row above shows what appear to be two triangular buildings and a pyramidic structure, but maybe these are just produced by shifting wind? What if we find roads?

Roads Anomaly

Over the tops of all types of hills and valleys, Venusian roads seems to go on forever which shows a reasonable probability that someone used those roads.

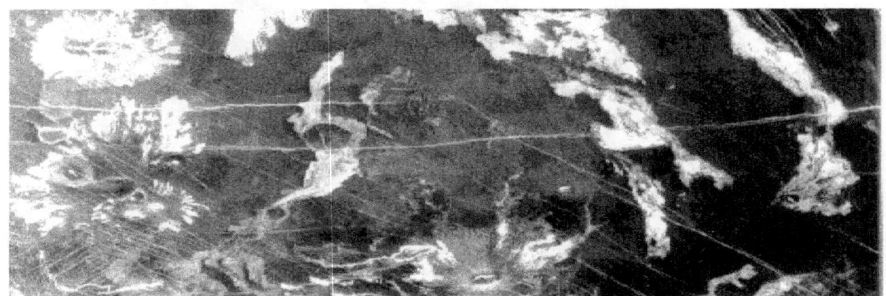

Below are a couple more similar roadways. IN the old days these probably went somewhere, but today they are just endless anomaly.

Roads-Over the tops of all types of hills and valleys, Venusian roads seems to go on forever which shows a reasonable probability that someone used those roads.

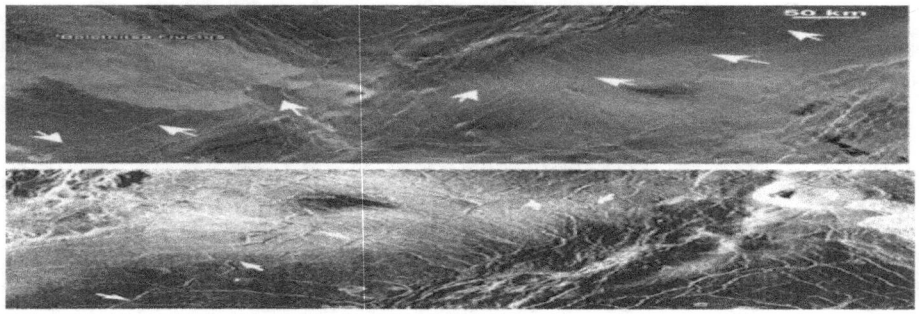

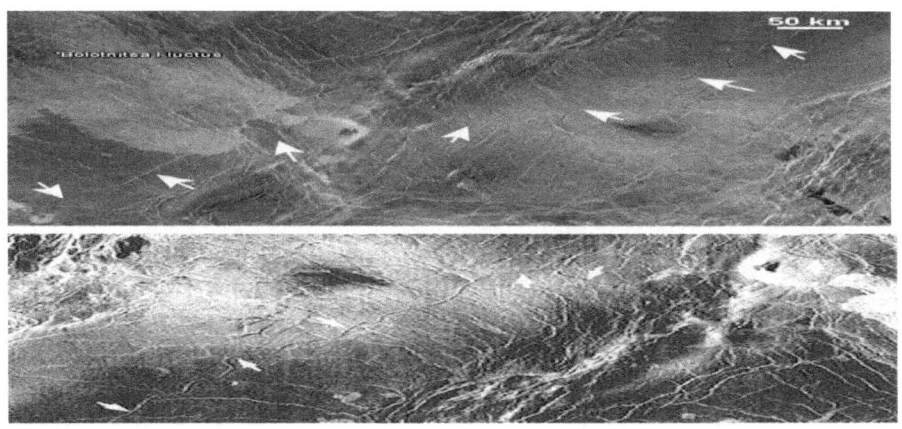

Remains of a City

It seems these roads possibly when to fairly large communities as this area seems to be laid out in a normal city grid pattern. In one of the protected valleys we even find what might be the remains of a city before being melted by intense heat and pressure.

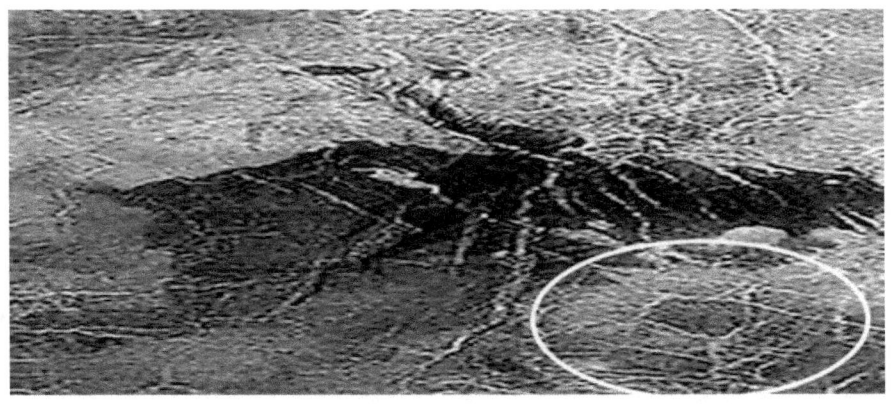

Venus War Anomalies

Before the atmosphere began to burn, everything must have been beautiful on Venus. Called the vain planet by the Bible writers, we can imagine that the additional heat of our sister planet made plants grow everywhere. Green and lush, our sister planet probably looked just like the Earth. Rather than living near the equator as we do on earth, the entire planet may have been livable and certainly the tundra areas would have been as green as our tropic areas.

Back on earth, the wars are really bad. The book Jasher indicated 1/3 of the inhabitants would die. They were getting worse and worse and there may have been continuous fighting for almost a thousand years. By now many people had moved underground. Soon, attacks on the planets and on the moon, were in fashion. In India, there is written evidence that battles were fought on the moon and in China descriptions of methods to transfer a detachment of men onto another planet has been found. The most likely planet would have been Venus. These descriptions and those below may paint a strong picture that the fighting was not confined to the Earth.

Indian Version-*Maharishi Bharadvaya*- In this work there are direct indications of gigantic battles in heaven.

Babylonian Version-In the "Epic of Etana" we read, "*Etana looked down and saw the Earth had become like a hill and the sea a well and so they flew for an hour and Etana looked down and the Earth was like a grinding stone and the sea* *like a pot. After the third hour, the Earth was only a speck of dust and the sea no longer seen"* [The ship, of course, was going into outer space.]

Chinese Version--Methodology of *how to send a detachment of men onto any planet* was described in ancient documents from Lhasa. These documents were found fairly recently and have been only partially deciphered. The remaining information is being deciphered as we speak, so we may find out more about the space war in the near future.

Greek Description-From Greek legends talking about battles between the gods we are told the following: *"Hot vapor lapped the titans, flames unspeakable rose bright to the upper air [outer space], lightning blinded their eyes."* [Apparently lightning weapons or weapons that affected plasmas were used in outer space.]

Bomb Blasts

Craters That Aren't Meteor Craters-This row of craters is not indicative of a meteor shower that would cause a random layout of variably sized blasts as the meteor exploded high in the atmosphere. These strikes are directed in a line on Venus. Here are seven blast areas in line. Someone was apparently trying to hit something during a strafing run of some kind. One thing that should be noted is that each of the blast areas is exactly the same size so the blasts could not have been random pieces of a meteor unless each piece came from the same source, all happened at the same times and all pieces were the exactly the same size and density. Even if one doesn't like to admit it, this is indicative of bomb blasts.

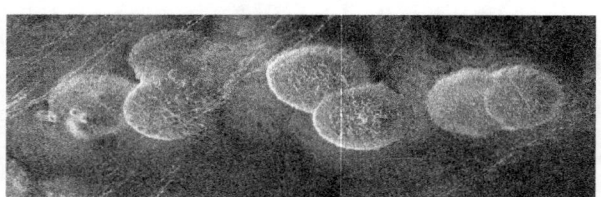 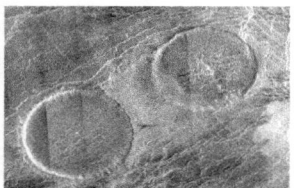

Volcanoes that aren't Volcanoes

To the right is a pair of perfectly round objects identified by supposed scientists as volcanoes on Venus. Has **<u>anyone</u>** ever seen a volcano as accurately cut and with two identical ones side by side before? I don't think so. I don't know exactly what they are, but at least now you are aware of them. Let's look at the first group as a good example of war. Here we find a city, Road, Bombs, and buildings. The next image will hopefully begin to bring out the truth.

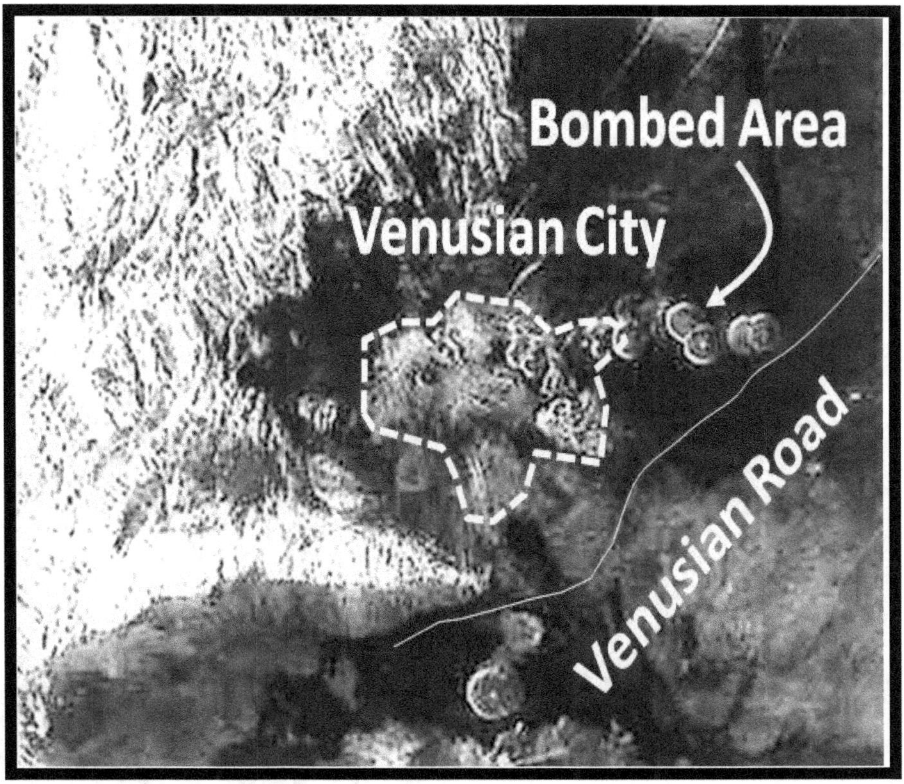

Let's look around the bomb blasts.

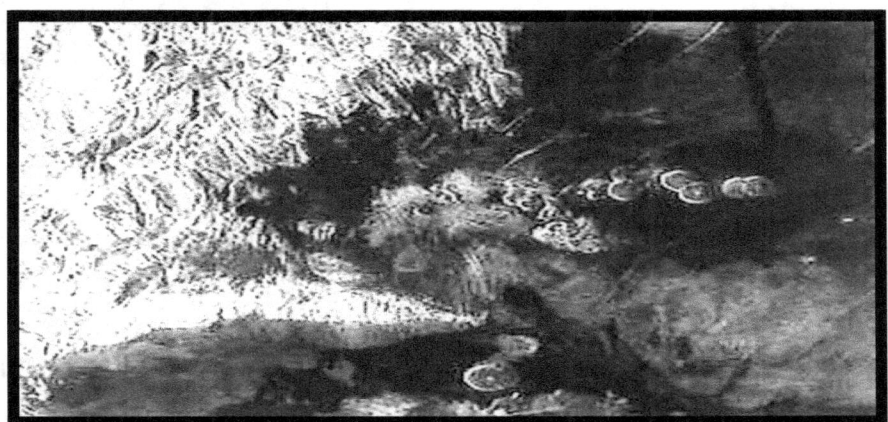

The image preceding is approximately 350 miles across shows smooth volcanic planes that border the eastern edge of the Alpha Regio Mountains. Within these dark plains we see some very strange pancake domes that are about 20 miles and the crater edges are about ½ mile high. That's not the unusual thing. While much of the planet's artifacts have melted away in the intense heat that is consuming the planet, some evidence remains to this day. Here is a blow up of the preceding image. Some try to ignore these features or come up with crazy explanations that don't make them say the war word, but hiding information does no one any good.

This row of IDENTICAL craters is not indicative of a meteor shower that would cause a random layout of variably sized blasts as the meteor exploded high in the atmosphere. These strikes are directed in a line on Venus. Here are seven identically sized blast areas in line. Someone was apparently trying to hit something during a strafing run of some kind. One thing that should be noted is that each of the blast areas is exactly the same size so the blasts could not have been random pieces of meteor unless each piece came from the same source, all happened at the same time, and all pieces were the exactly the same size and density. This is indicative of bomb blasts not meteors or volcanoes. Please notice, also that the blasts are along a very distinct roadway and if you look closely it appears that they hit some objects that were rectangular as the blow-up shows where 2 of the blasts struck buildings. I put in a few bomb blasts from bomb runs during World War II next, but these are on Earth.

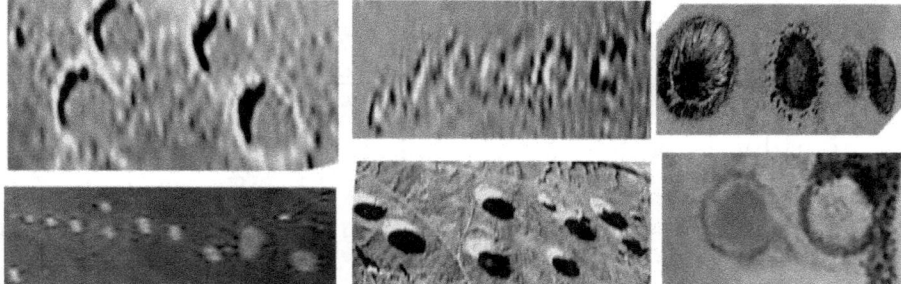

Venus Battles- By accounts in the Bible and other places, there evidently was a very active group of warriors lived on Venus so we can expect it had its share of attacks. The Bible and the book of Jasher tell us of a second war to try to take "control of heaven". As if the losers of the 1ˢᵗ Heaven War described in the Biblical testimony, that occurred 120 thousand years ago had station troops on Venus to Try Again! According to several texts, Satan had stationed men on the outpost known as RAHAB and God decided to halt this meager attempt by destroying the entire planet. I don't know what that means exactly, but what is apparent is that you should never get God made at you. He destroys planets like you would destroy a lump of sugar

Gnostic Descriptions- This comes from the book *"Origins of the World'-Before the consummation of the age, the <u>whole place will shake with great thundering</u>. Then the kings*

will be intoxicated with the <u>fiery sword</u>, and <u>they will wage war</u> against one another. Then the seas will be disturbed by those wars. Then the sun will become dark, and the moon will cause its light to cease. The <u>stars of the sky</u> [Venus?] will cancel their circuits. And a great clap of thunder will come out of a great force that is above all the forces of chaos and <u>their heavens will fall one upon the next and their forces will be consumed by fire.</u> Their <u>eternal realms, too, will be overturned and his heaven will fall and break in two</u>. Everything is good but-"only to be replaced by <u>the spawn of a lower star.</u> Over the world <u>broke the great waters, drowning and sinking, changing the Earth's balance</u> [Before the worldwide flood, the Earth had huge Earthquakes and Civil Wars that turned into Worldwide War. Venus sent meteors from the sky. The light of the sun and moon was darkened by the incident; meteors destroyed and Venus or its moon split in two. Everything was great but all of a sudden, Venus caused massive floods, and then the earth axis was changed.]

Tibetan Description- this comes from the book **"Book of Dzyan"**. The <u>Guardian of Venus</u>, with a mighty roar of swift decent from incalculable height was <u>surrounded by blazing masses of fire which filled the sky with shooting tongues of flame.</u> The vessels of the lords of <u>the flame flashed through the aerial spaces. Stars showered down</u> on the black-faced while they slept. A large heavenly stone crashed into the Earth and caused a long time of darkness and cold. Thus did the <u>large, lumbering creatures come to the end of their season. The first great waters came and swallowed the 7 great islands.</u> The <u>holy were the only saved</u>. The unholy and most huge animals were destroyed. Few men remained <u>The guardian of **Venus came down to Earth** with his disciples in</u>

huge ships. The king of the Dazzling face, sent his air-vehicles, with pious men inside, to all the other chiefs saying, "Cross the land while dry. The lord of the storm is approaching." Let the dazzling faces take the flying ships away from the lords of the dark skinned. The waters had already moved, but the nations had now crossed the dry land and were led to the lands of Fire and metal. The waters covered the whole world. [This is a great description. There were many explosions on Venus. The explosions fostered many meteorites, which filled the earth's sky. Meteors from Venus hit the earth and it was followed by a shift in our axis. This is saying the remade dinosaurs died---we can be pretty sure none were in the Ark of Noah. Some were warned of the great catastrophe. The giants were destroyed and some survivors from Venus came back just before Venus turned into an inferno. Some of the ancient people escaped the doom of the worldwide flood by using the flying ships that I mentioned before.]

Brazilian Description- This comes the a "history by the Mongulala people- *The Blood Age was the beginning of the Mongulala history. It started immediately after the Golden Age. Critical information about the events of the era was written on animal skins. In the west, where now is only water there was a large island. A second gigantic mass of land was in the northern part of the ocean as well. Both lands were buried under an enormous tidal wave during the first Great Catastrophe. It occurred at the end of the war between the two divine races. The war between the two divine races did not only lay waste to the earth, but also to the worlds of Mars and Venus.*

Biblical Tests- This comes from **Genesis 6-** *And it came to pass, when men began to multiply on the face of the earth,* *and daughters were born unto them that the sons of God* [These were called Nephadim people]*saw the daughters of men that they were fair; and they took them wives of all which they chose.-- There were giants* [These were the Anakim people]*in the earth in those days; and also after that, when the sons of God* [Nephadim People again] *came in unto the daughters of men, and they bare children to them, the same became Giants which were of old, men of renown. And God saw that the wickedness of man was great in the earth, and that every imagination of the thoughts of his heart was only evil continually. And it repented the* LORD *that he had made man on the earth, and it grieved him at his heart. And the* LORD *said, I will destroy man whom I have created from the face of the earth; both man, and beast, and the creeping thing, and the fowls of the air; for it repenteth me that I have made them.-- The <u>earth also was corrupt before God, and the earth was filled with violence</u>* [Massive wars in the Pleistocene Age] *And God looked upon the earth, and, behold, it was corrupt; for <u>all flesh</u> had corrupted his way upon the earth.*[Animals were corrupted by genetics] *And God said unto Noah, The end of all flesh is come before me; for the earth is filled with violence through them; and, behold, <u>I will destroy them with the earth</u>.*

If there was a war, there must still be signs of life, you might think.

Venus Life Anomaly

All I can say is three things- Sulphur-Dioxide, Carbon-monoxide, and Ozone. Using data from the Russian Venera space missions and also the US Pioneer Venus and Magellan probes, researchers studying the high concentration of water droplets in the Venusian clouds found hydrogen sulfide and Sulphur dioxide. These two gases react with each other, so they should not be seen in the same place unless something is producing them. Despite solar radiation and lightning - the atmosphere contains hardly any carbon monoxide. This suggests **something is removing the gas**.

Ozone Indicates Current Life on Venus

One belief is that bugs living in the Venusian clouds could be combining sulfur dioxide with carbon monoxide and possibly hydrogen sulfide or carbonyl sulfide in a metabolism similar to that of some early Earth bugs. The Venus Express [2008 to 2012] found Ozone in 2011. This doesn't make sense in that it has only previously been detected in the atmospheres of Earth and Mars. On Earth, it is of fundamental importance to life because it absorbs much of the Sun's harmful ultraviolet rays. Not only that, it is thought to have been generated by life itself in the first place.

Argon Gas Dates the End of Civilization

The main curiosity, however, found by the Magellan probe was that the atmosphere contains high levels of the isotopes of argon, neon and noble gases. These high concentrations of noble gases could only mean that the current atmosphere of Venus is extremely young, because noble gases don't combine with other materials and escape easily into space; even with a thick atmosphere. Whatever killed most of the life happened not very long ago.

Worm Thing

The Russian Veruna landed and took a number of pictures before it was finally destroyed by the intense heat, but they seemed to have photographed something strange. Images up until about 90 minutes they just saw the ground, but look at the image taken 100 minutes after landing. Something came out of the ground and disappeared in the 113-minute image.

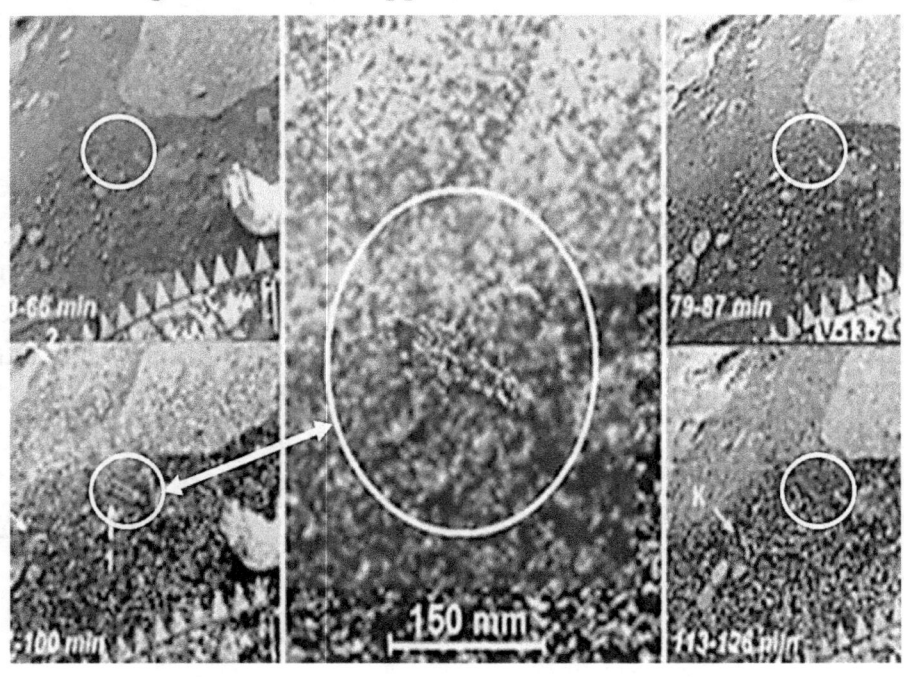

Algae Anomaly

A fairly recent meteor that his around the time of the Carolina Bays, is very interesting in that it contains Algae. We must recognize that left in space for too long would erode such a find, so the meteor came for a close location, such as Venus. This may show that the rivers and oceans of Venus were filled with life. The images of the algae and meteor are shown below.

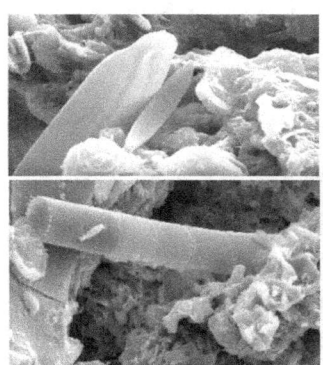

6000 years old algae in meteorite

Flytrap Life

As I mentioned before the strange microcosm of Venus flytrap plants only in the central location of the Carolina Bays makes everyone wonder if the plants were not indigenous to our planet and fell with the Meteors. If so it may be a good representation of plant-life on Venus before the meltdown. I'm not saying all the plants would eat flesh.

Mars Colonization Anomaly

Old Mars Anomaly

> *We really need to go on to Mars as it has played a major part in our civilization but like Lunar and Venus, the influence has been hidden away in a neat little anomaly package so no one would question how our history books have been laid out.*

Not only was it most likely responsible for the massive hole in our planet we know as the Pacific Ocean and many of the mountain ranges that cover our world, but also, we need to exam possibilities of colonization, and even terraforming. I'm not talking about visitation from Alpha Centauri, I'm talking about regular people again. This not only affects us here, the large group of adventurous re-colonizers building for a near time blast off can gain solace in these details. Our historians have been so kind that most people believe if someone is on Mars they are from another world or simply Martians when our ancient historians and physical evidence tells us colonization on Mars was a long time ago and it was by humans.

Interest in Mars being inhabited by Martians really started in modern times about 150 years ago. Let's get an understand of that before going on. Unlike Venus, there are no indications of Mars in our Bible, but that doesn't mean that in our distant past, people lived on a much different Mars.

1877 Schiaparelli [Canali]-It was then when Italian astronomer Giovanni Schiaparelli saw deep trenches meandering across the red planet's surface, which he called "canali." This meant "grooves", but someone thought he was saying "canals" and it was believed if canals were on the planet, Martians were as well. Not many believed they were people because it would mess up what was determined to be "True" history.

Lowell [Martian Life]-The American astronomer Percival Lowell continued Schiaparelli's work, but his idea was to popularize the idea that Mars held life. His enthusiastic interpretation of the canals as Martian constructions alienated his assistants and annoyed Schiaparelli himself. Below are his detailed drawings of the deep canal like indications he "saw".

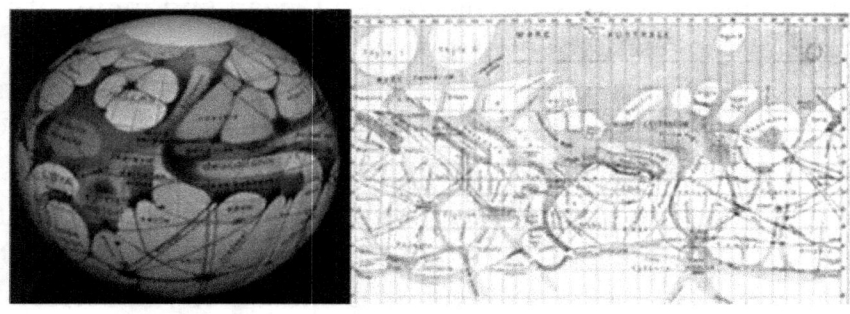

1898 [Article]-An 1898 article in *"The Atlantic Monthly"* noted that Mars might be in *"an advanced stage of evolution"* compared to Earth. We were now certain of the advanced Martians; we simply didn't know what their intensions were.

1906 [Book]-In 1906, Lowell published a popular book, *Mars and Its Canals*, proposing that the canals served to transport water from the poles to the planet's more arid central plains. They were proof of Martians being civilized and advanced.

1907 [A.R. Wallace]- In 1907, *Is Mars Habitable?* by A.R. Wallace directly critiqued Lowell's book, and presented Natural reasons for the apparent indentations.

1938- [Orson Wells]- To keep people from hysteria, Orson Welles developed *War of the Worlds.* Wait a minute! He did the opposite thing. According to his broadcast, the canals had been signs of a more developed, and hostile group of metallic Martians. Mars had become a dry, red, craggy landscape populated only by nasty Marians bent on our destruction.

A Truer Overview of Mars

What we are going to review are many, many seemingly anomalous articles and characteristics found on Mars. As simply calling them anomalies is not a reasonable answer many times, we must look for other plausible considerations. To really understand Mars and the potential for ancient settlers, we need to look at a time before the Pleistocene Extinction when rivers, and oceans, and "Air" were still available. The ancient atmosphere also allowed the ground to hold more heat. It seems people have almost always been at war and during the last part of the Cretaceous and towards the last part of the Pleistocene, warring was especially horrible and we can believe some colonists either left to separate from the endless violence, or people were stationed at remote outpost for some tactical advantage.   What they found was a fairly cold place, but plants still thrived. Most would have built structures low to the ground or even under the ground where possible.

The problem was that Mars was and still is spinning too quickly to hold on to any substantial part of its atmosphere. Each year there was less and less. This would have taken many thousands of years to deplete, but soon underground dwellings also had to provide oxygen as the Martian atmosphere was losing more oxygen than could ever be replenished and over a long time, the planet all but died.

Strange Synchronization- With only half the circumference of earth or about 1/10th the volume, Mars currently has a day that is almost identical to the current earth day. As our spins have somehow been synchronized, one might think the earth had something to do with the high-rotational speed of Mars. This high-speed spin may be an effect of it losing half its crustal material after a nearby contact with Earth, 400 thousand years ago. Given the much lower gravitational pull of the planet, this may be too fast to hold onto an atmosphere and water like that found on Earth so why is there so many indications of water? Hint! Mars used to have a massive amount of water, rivers, oceans and the like.

Unusually Small-Mars is currently much less massive than any other planet, except Mercury and Pluto which have been generally considered to have been escaped moons. People wonder why it is located where it is. As we have already discussed, this suggests Mars once was substantially larger and was broken in half.

No Northern Craters-The Martian southern hemisphere is peppered with craters. Few are seen in the north as if half the planet is gone. This crustal dichotomy is almost a perfect circle. This suggests that an extremely close cataclysm almost ripped the planet in half.

No Crust in the North- The smooth crust of the northern hemisphere is only about a kilometer thick, compared to 20 kilometers in the south. This indicates that the northern half of the planet has been ripped away. While this fact is unmistakable, some choose to ignore it.

Too Few Asteroids- The nearby asteroid belt is clearly the remains of a planetary mass that once was spinning around the sun. The problem is there isn't nearly enough mass to have been an entire planet as if the massive amount of crustal matter that was sucked into space from a near collision of Mars and Earth that formed our Pacific Ocean and peeled away half the crustal surface of Mars, left the asteroid debris. I know it doesn't show up in your history book, but where did the asteroids come from, where is the other half of Mars and how could a super continent of Pangea form on only one side of our planet?

Terraforming for Air-Today there are signs of terraforming and some level of civilization may still in place. For that possibility, we will investigate secret American endeavors called and the Rainbow Project and the Pegasus Project. With such a large number of people claiming to be part of this mysterious event, we cannot ignore the probability of life still hanging on at the output we call Mars.

Water on Mars-For a long time the concerns over Martians faded as scientists found no water to support life, but recent news of liquid salt water on the planet's surface has people wondering again about the Martians and what they might want with us. Whether texts books tell us or not, hysteria over Mars has reached a breaking point over the past couple of years as NASA's "Curiosity Rover" beamed back breathtaking photos to Earth. As the rover now is being eaten by something being picked up by the rover's track, even more questions are coming about. Also, a number of flying objects have been viewed which makes someone think someone is flying the flying object. Another rover will be sent up in 2020 to find more about life on Mars. Fist let's get the water concerns out of the anomaly category.

Martian Water Anomaly

Something happened on Mars that destroyed everything and soon the atmosphere left. That event was not the near collision with earth that left half of the planet blown away and the remainder pitted with craters. That earlier event, most likely occurred about 400 thousand years ago. Even after that catastrophe, water was still abundant on Mars and with water there was, most likely a reasonable atmosphere. At a later date, the atmosphere began to thin and the water began to leave. Although we don't know the exact time period for this final disaster for any colonists, the picture following, left, shows the remains of a huge sea on Mars. Please see that the remains are not very old and the details are still distinct, so we can be fairly certain that it was not very long ago. Some researchers place the rippling of the ground at about 17 thousand years ago by its strong definition. Next to that image is a portion of a river that still appears to have a little water remaining.

 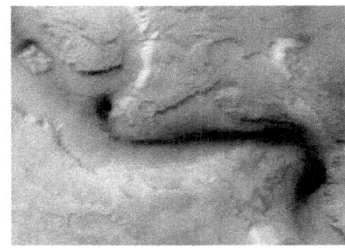

Speaking of the rippling of the ground and a 17 thousand year ago loss of water, here is another twist. From studies of Martian meteoric material, Dr. Leshin of the University of Arizona and others have concluded that Martian water originally contained higher deuterium levels than previously thought which means that the Martian atmosphere has lost ½ the amount of water through the eons than "dry planet" models suggest. Therefore, some have concluded, there must be a huge ocean-like reservoir of water beneath the planet's surface. Not 17,000 years ago, I'm talking about water on Mars today. The recent picture, below left, shows a reflection off of a shiny surface. It also appears to be a frozen lake on Mars and that brings us to terraces.

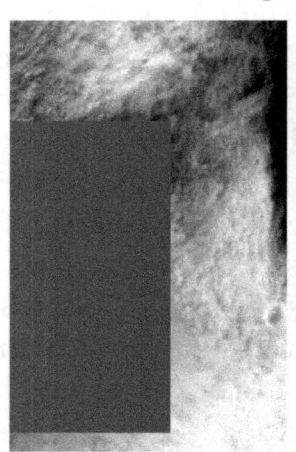
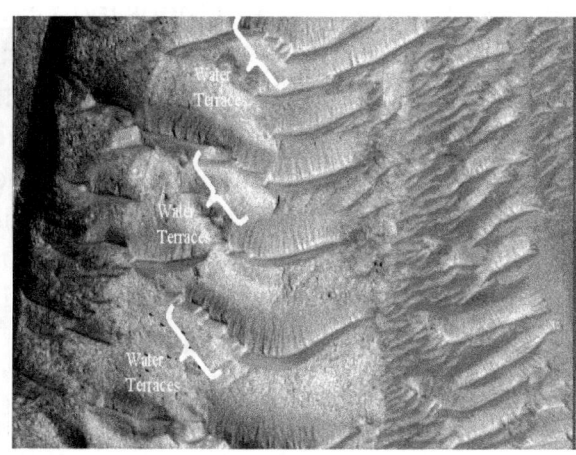

Water Terraces- People who lived on Mars, evidently, build water terraces to harness and use the water before the atmosphere got too thin and the water left the surface. [See right above] When we blow up one of the sections the details show a get capability in building these terraces and controlling water flow.

Even without the terraces, we see that water is so abundant underground that is seeps out without drilling. The three images below were taken over a few months' time period. Notice how the darker wet areas change position and size.

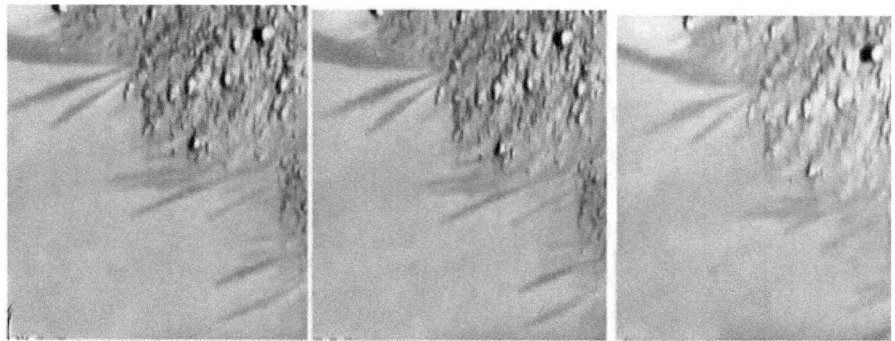

The images below are more of the wet areas showing water seeping out , naturally,

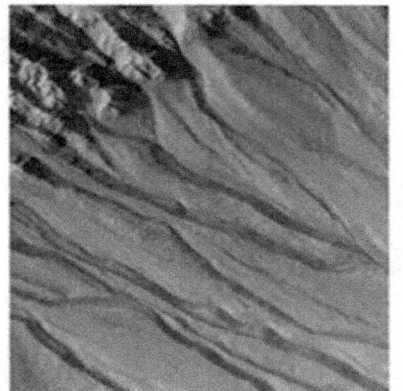 

Still More Seeping Water-Not to get you overrun with water images, these are the last water seeping images. Even today We know that running water can be found on the surface of the planet. Here are a couple of timed photos that prove it.

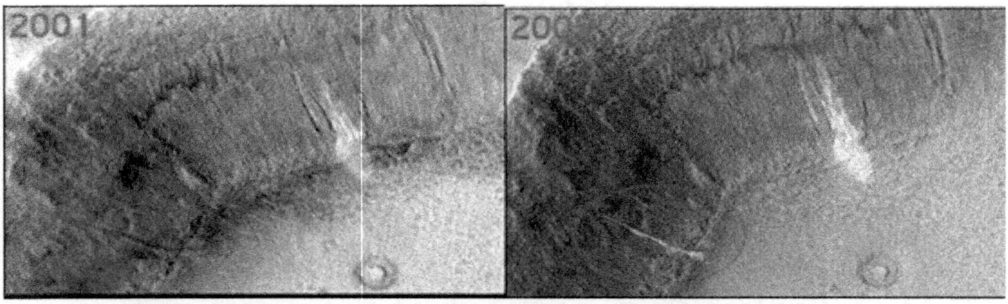

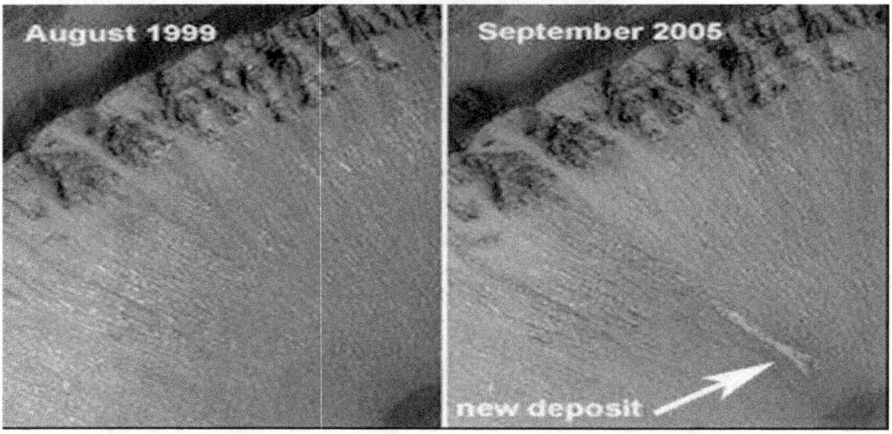

Where there is water there is or could have been life so the next image is important as surface water is puddled in a number of ancient lakes as shown.

Rivers-A good example of a dry river on Mars is shown below. There can be little doubt that the water flowed along this path in the not too distant past.

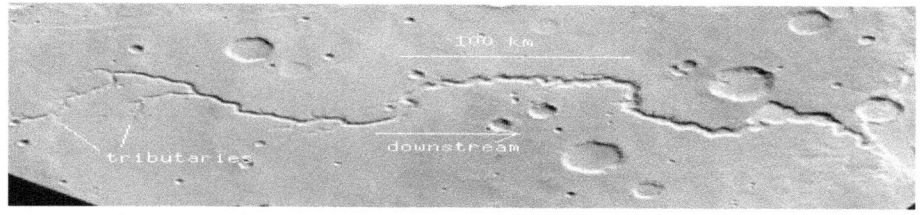

As water reached the open sea, a massive river delta formed. The remains are shown next.

The following image is the remains of another massive river delta no longer flowing with any more than the tiniest bit of water, but, at least, it's something and these are still distinct showing the water left in the not too distant past.

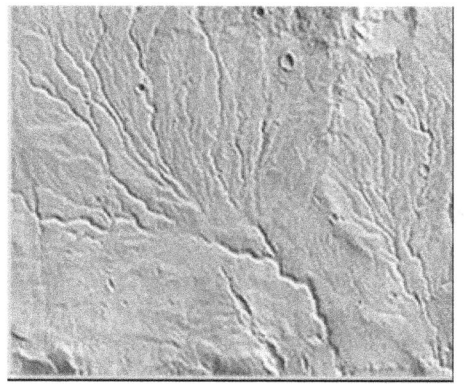 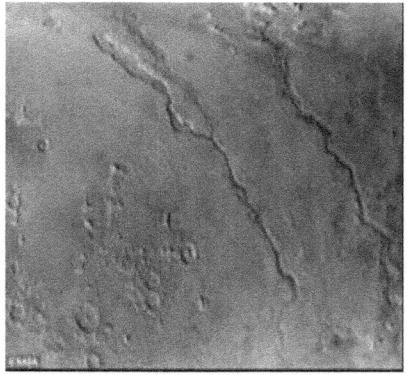

This last river picture shows that the rivers are ancient and older than this meteor strike that closed off the waterway.

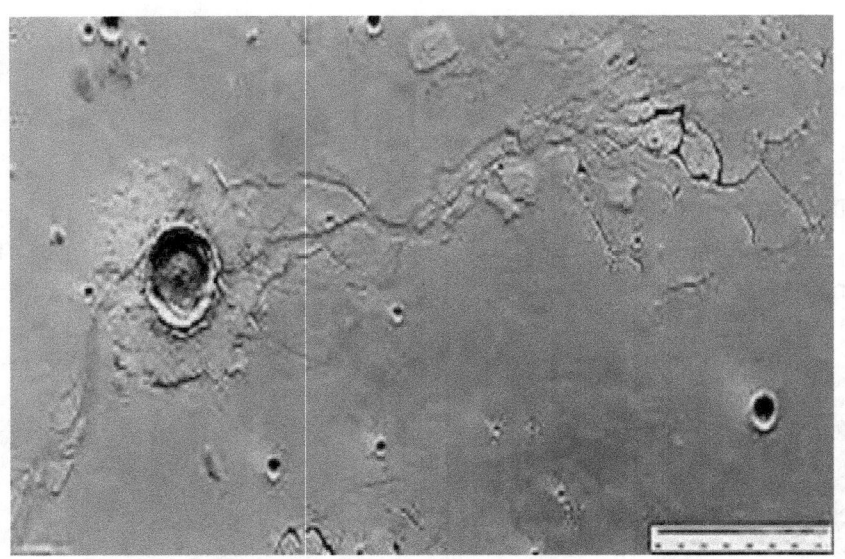

So, we had and still have water. Now the question might be; what if we find buildings and cities?

Industrial Anomalies

There are many signs of life on Mars and many anomalies that strongly suggest that this was human life not homegrown Martians. Possibly, a majority of the population was destroyed during eve earlier fighting, but surely some remnant survived. This goes along with both scientific reason and religious record. Besides the massive complete cities that are evident, there have been found an enormous number of artifacts that tell us that life on Mars has only been extinct for a very short time.

The above NASA image has been enhanced below to show what appears to be buildings, streets, and all the rest in this protected valley.

If that was all, we might be wondering, but here is another and another on Google Mars.

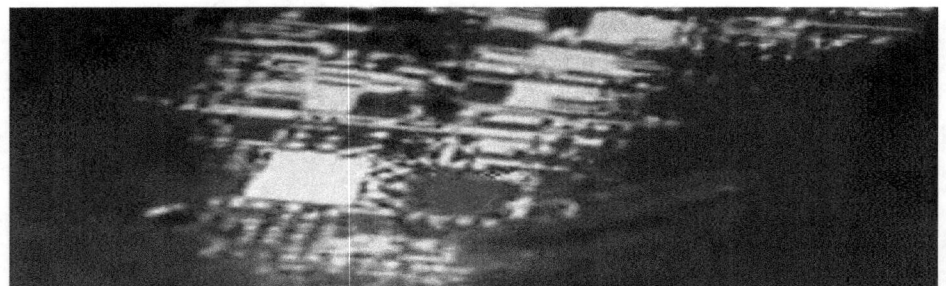

At another location, we see this! Again, in a protected valley, this industrial area is hidden next to massive cliffs. We find buildings streets that look like a city. Next to it is still another ancient industrial center [Below right].

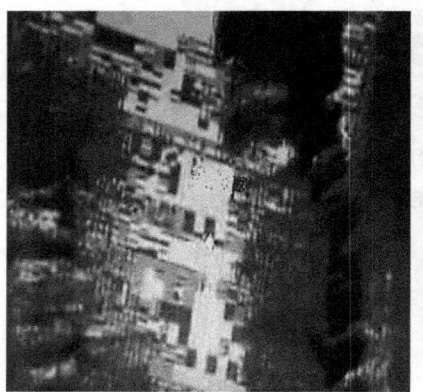 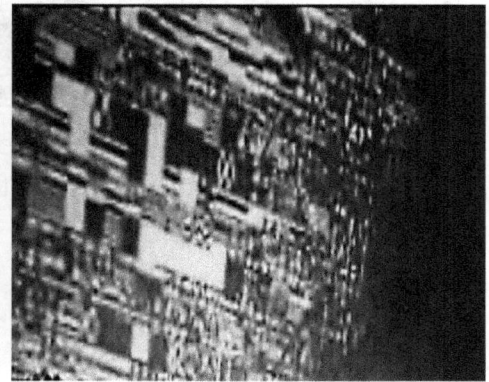

Another Huge Industrial Area- I know what you are thinking these are doctored pictures, but they are not. Most come directly off Google Mars from images taken from our many satellite trips. For this section, I want to concentrate on a section of Mars near the municipal area of Cydonia. The next image is a picture from near that area. In this section of the countryside, we find the best examples of an industrialized planet with its factories and closely defined structures. I circled a couple of areas to focus in on, but there can be little doubt that this area was laid out to be similar to our modern industrialized areas.

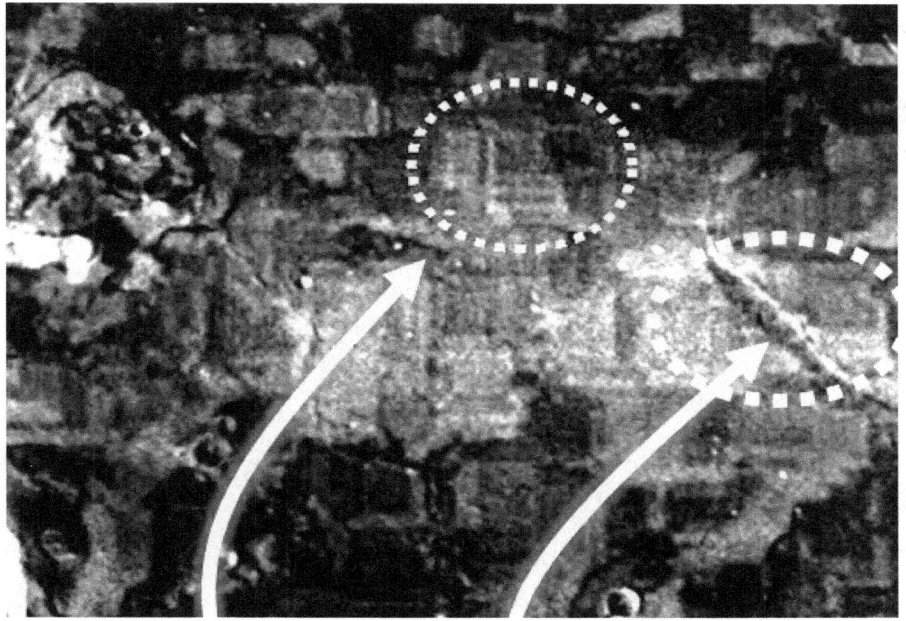

If we look about the middle top portion of the image [the part I circled], an industrial site emerges. In the following picture, the inserts for windows or whatever can be easily made out. These are not natural window cutouts of a rectangular cliff, but instead they show characteristics of being manmade.

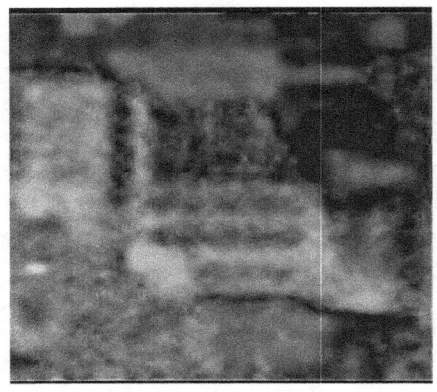 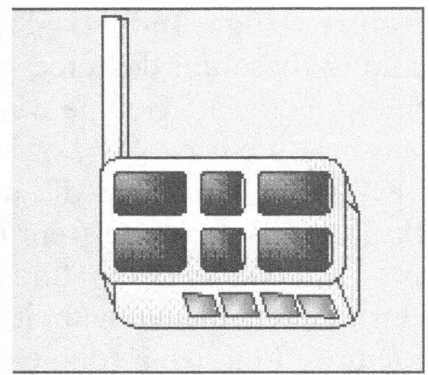

You can see the big building to the left and the windowed building in the middle along with the rectangular flat surface along the top and some type of grid work connecting the flat surface to the windowed building. I drew the windowed building to the right. If we look at the same picture and blow up the middle right section, we see more.

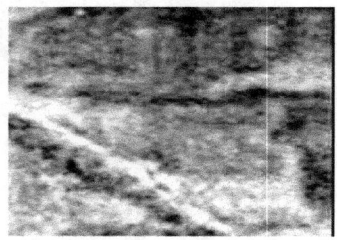 

Except for the gash down the middle of the town, this could be some earthly business area.

Still Another Industrialized Section-This comes from an unassuming area known as Hale Crater. The tiny square on the picture below shows the section of the protected crater that will be blown up for viewing. If we blow up this section we find that the above details are certainly not the only indication that Mars once was a hugely industrialized planet. Details are shown below right. Does this look like an industrial section or a dead planet?

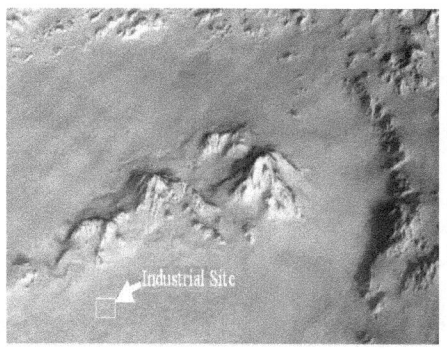 

If we blow up the section on the left, we can see all sorts of buildings with rectangular gridwork and parallel walls, and regularly spaced building separations. There can be little doubt of the massive structures in this area. [Abover right]

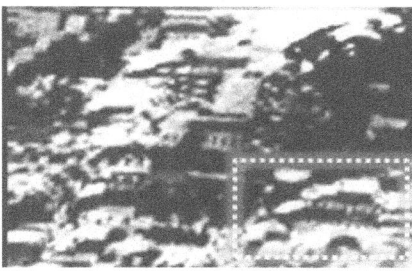 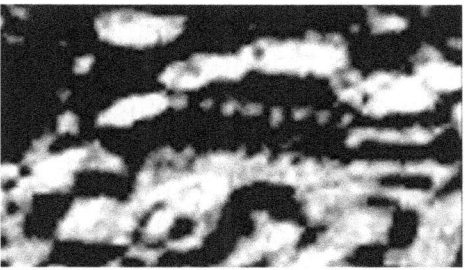

If we blow it up more we can view one of the most magnificent cultural centers on the planet. Notice the eight visible columns in the front and the multilevel terraces, the various pools [now empty]and round silo looking water pressure tank or similar functioning tank. These all appear to be fairly modern which gives us one level of understanding concerning colonization in the no too distant pasts, but we also find ancient looking cities similar to those we have found by the ancient preMaya.

Ancient Cities Anomaly

Almost no self-respecting historian would ever say what is being seen on Mars is reality, but this next set even could be more objectionable as it shows colonization many thousands of year ago. All the preceding industrial area were found to be protected by mountainous structures around the complex, but we also find geometrically laid out areas out in the open. Besides being out in the open, these "cities" appear to be thousands of years older and filled with Martian dust. Many of the structures in this group do not have straight walls. Instead; many slope, like pyramid structures found all over the world. Before I get into this section, let me say that my pictures do not do justice to the actual finding of the various researchers on this topic. My desire here is to let you know about the findings and provide an overview. There are at least 7 areas on Mars being studied today that look like they were some type of city with massive structures. The one below is one of the "cities".

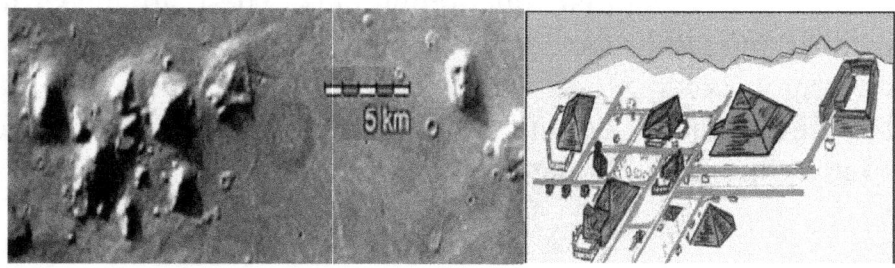

Called Cydonia, this "city" contains pyramids, a high walled fort, whose walls have collapsed centuries ago, and outside the city, we find a curious face-like carving on a hilltop about 10 kilometers north of the city limits. Literally dozens of well-defined straight edges, right angles, and geometric forms don't come together in a single place naturally so many now believe that someone put them there. Again, I don't mean little green men, I men regular men from earth. More details concerning Martian habitation can be readily found through many sources with one of the premier investigators being Richard Hoagland, so don't just look at this brief overview and think that this is the only evidence of human existence on Mars. I am only presenting a small quantity of elements just to show that there is enough evidence to even allow the most ardent "non-planetary habitationalist" cause to wonder.

Martian Town #2-Like the first city, this second "city", shown to the right, contains many well defined pyramidal "buildings" that are positioned in a matrix that resembles a city.

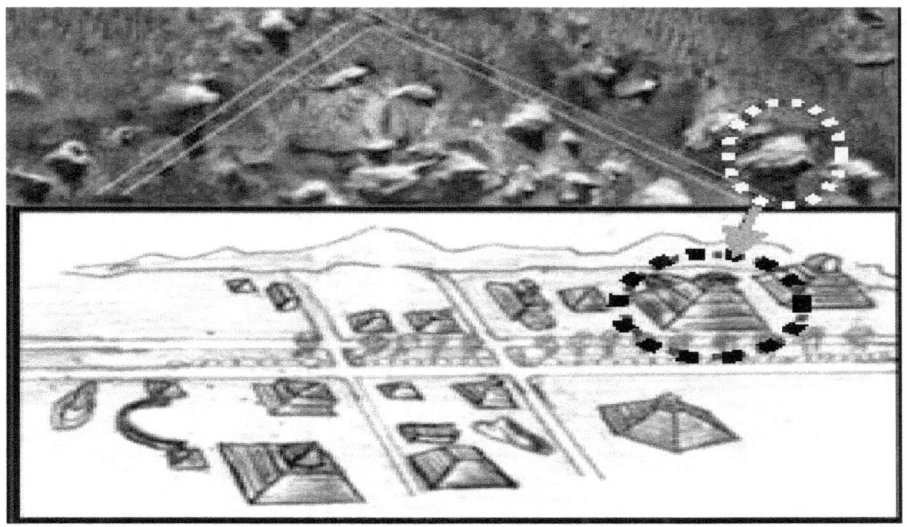

The pyramid on the right is especially interesting in that, even the steps up to the building can be made out. The drawing to the right may bring out some of the features. Please remember, there are better pictures of this area and the other areas shown if you check out works by other researchers. My mission here is to simply bring up the probability and show the right angles, the building like structures, the groupings of structures, the apparent roadways, the geometric similarities to earth structures and towns, and allow you to understand that the planets have very apparent signs of life.

Martian Town #3- The third "city" below shows the effects of some terrible ripping on Mars as a huge crevice splits one of the cities in two. Also, a crater can be seen in the city itself with a large building almost to its rim, showing that the city may be older than the crater. Right angles can be made out everywhere in the city area, which is not typical of non-manmade structures. Look especially at the shadows to examine details. There can be little doubt that this array of structures must have been placed here many thousands of years ago. Probably during the early Tertiary years.

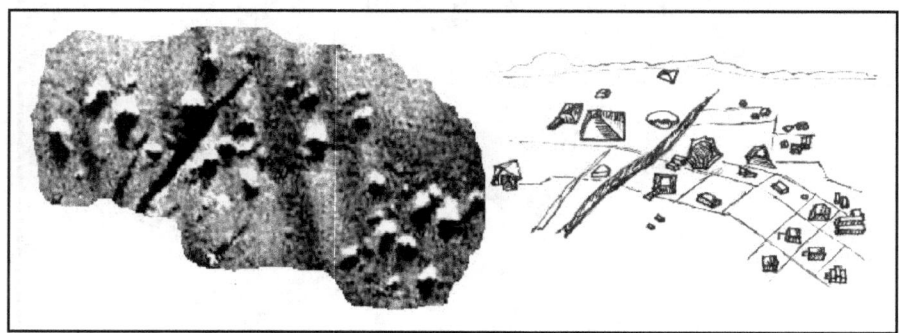

Martian Buildings Anomaly

Besides industrial areas and ancient towns, we find all types of other structures on the Martian surface. Here are just a few.

Stepped Pyramid-The first is called the stepped pyramid for good cause, it is stepped as shown below left. I have drawn its general shape to the right for clarity. Several steps and platforms are clearly visible. Like the pyramids found in Guatemala, several stepped pyramids have been found on Mars. This one is still in pretty good shape considering the time it must have been made. [following right]

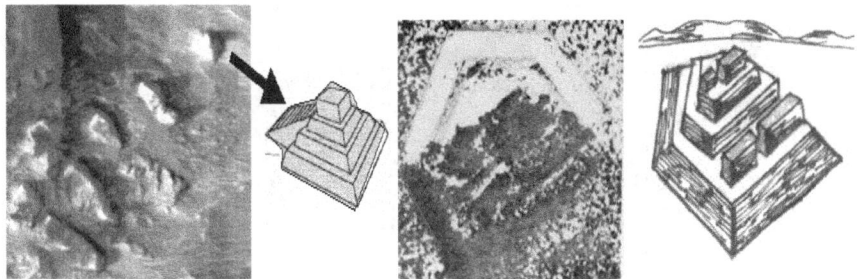

Pentagon-Like our own pentagon, this building [Above right] certainly describes the greatness that once was had on the planet Mars.

The Famous Martian Airport-If those photographs are not enough to convince you; look the picture from Mariner 9 that

Richard Hoagland termed as the Martian airport-out by itself, as you would expect, but with unbelievable detail. The, Tom Penner, rendering of the complex certainly shows that, at one time, this was an airport or huge shopping center or something else, but it was not an accidental structure. Actually, the structure is mostly underground rather than above ground as the artist conception shows, but you can get the idea just the same.

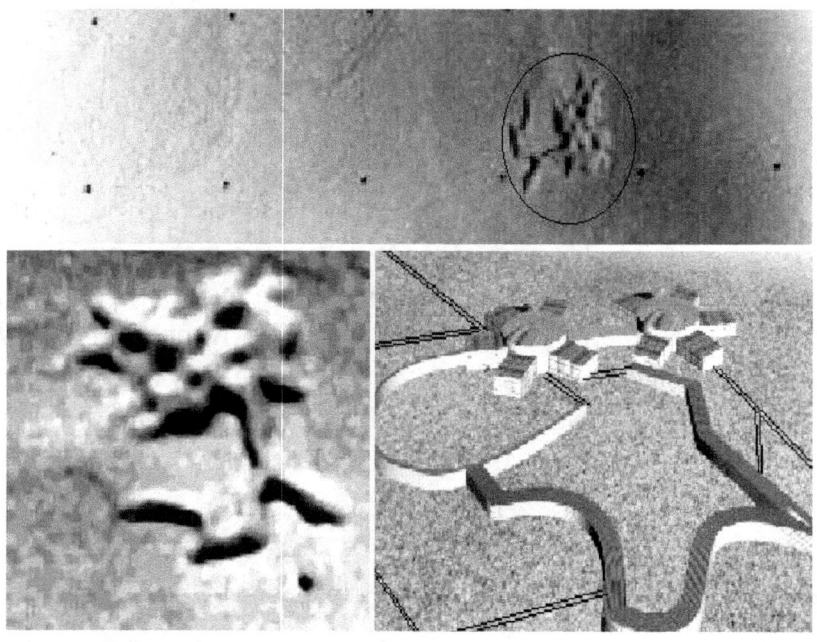

We could go on and on looking for more signs of human occupation in this section, so I think I will.

Martian Sea Port-The picture following shows what has been reported to be a Martian Seaport. This area is commonly known as the port or seaside retreat. Note the extremely regular shapes associated with the building on the cliff that once could have been adjacent to a waterway below. I have superimposed a drawing of the building for clarity. The right-angled surfaces of the building are still very distinct showing that it was not abandoned extremely long ago. I know the three squares with round areas on the center of each looks like some odd-looking home, but think of the guy that owned this retreat as an eccentric. This area is commonly known as the port. Note the extremely regular shapes associated with the building on the cliff that once could have been adjacent to a waterway below. I have superimposed a drawing of the building for clarity. The right-angled surfaces of the building are still very distinct showing that it was not abandoned extremely long ago. To the right is shown the area where the Seaport was found.

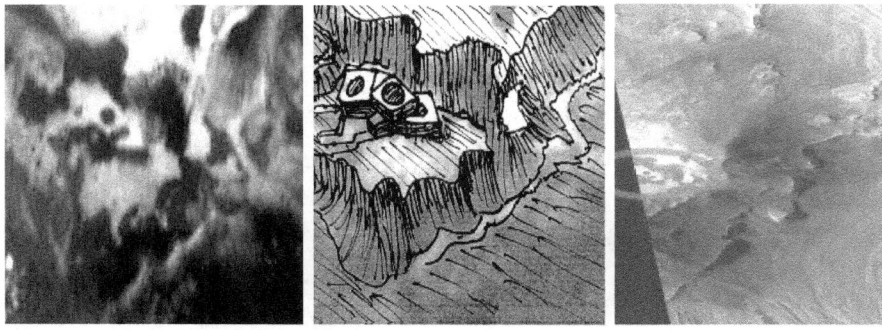

Town of the Shore- Still another set of remains looks like a city once was along the seashore.

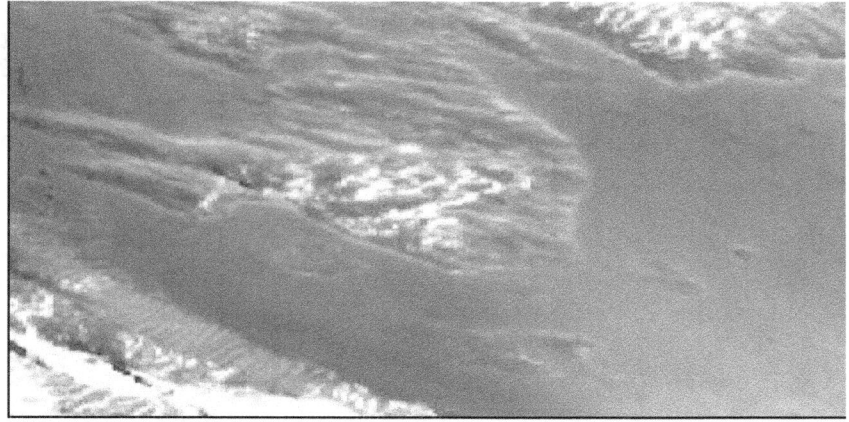

Walls- Other places have been walled to protect areas. Soon it was evident they could not be protected and Mars would have been abandoned. The massive wall below could not have been made naturally.

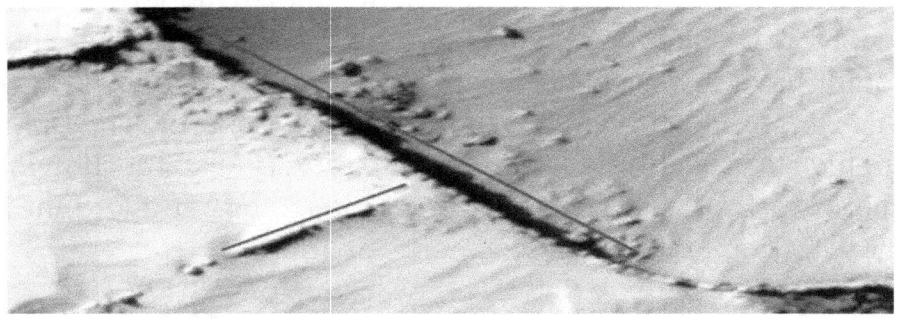

Amphitheater-The following image seems like some great amphitheater with buildings around the one end, possibly as an ancient portion of a town, but the massive curved structure with a square stage in the center is hard to ignore.

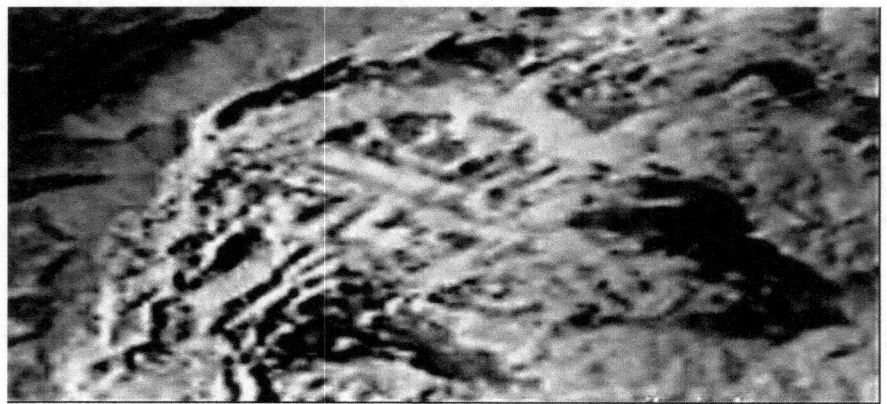

Tower Anomalies

Just like on the moon, Mars is full of towers. Here are just a few of them. In the next image, we can see a major tower still standing. This one seems to have a base complex below with long corridors to the huge tower. Protected from the environment by the nearby rock face, it has stood the ravages of time fairly nicely.

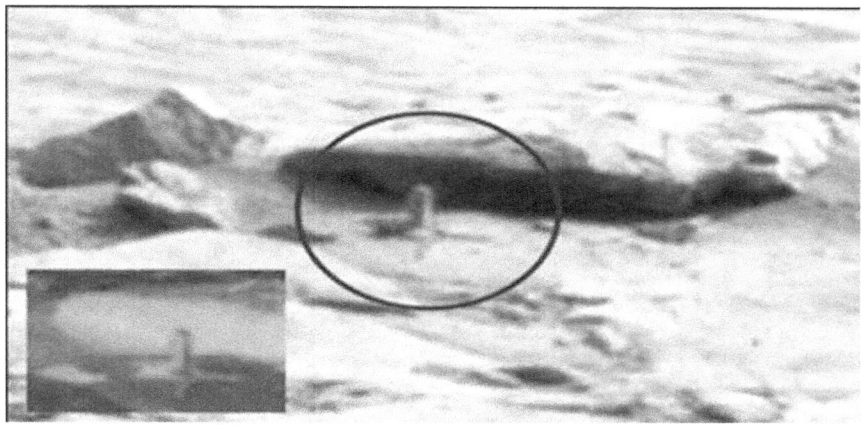

Next are two more. Interestingly, the first one is so smooth, it acts like a beacon for ships that will never sail.

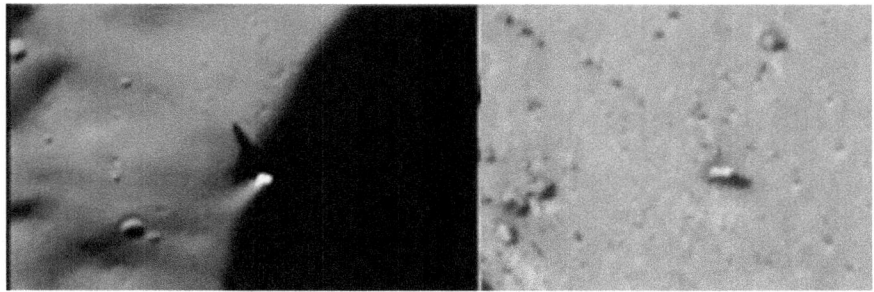

Three in a row.

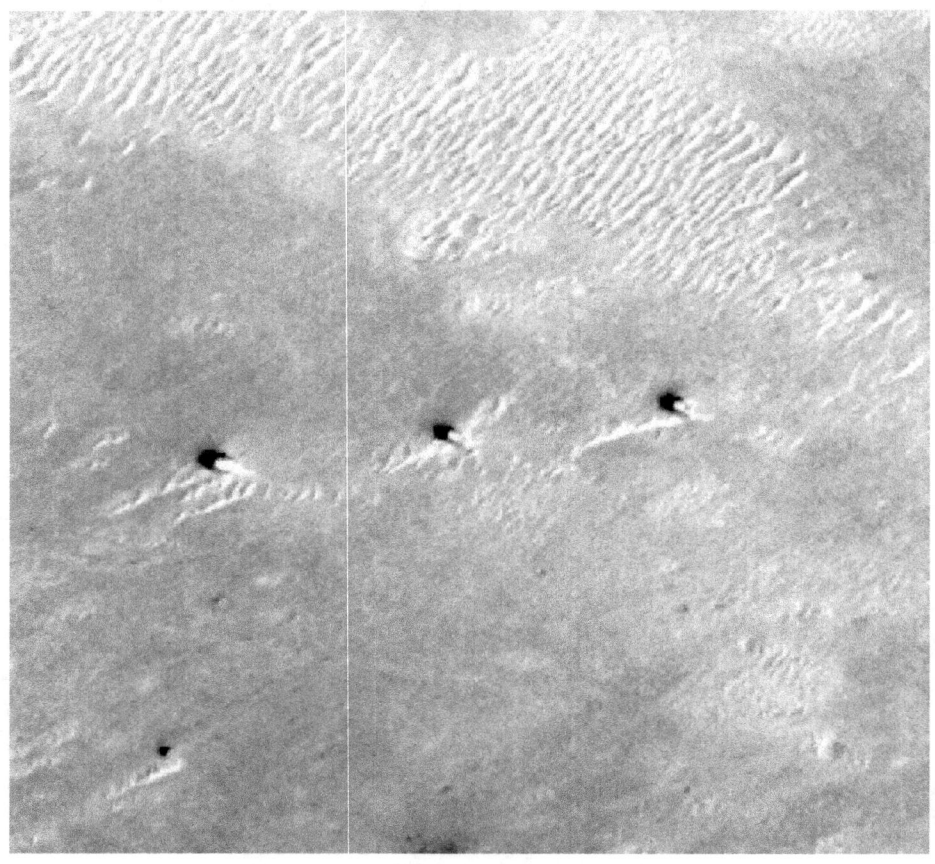

Martian Road Anomalies

Certainly, someone would believe colonists were once on or even still on Mars if we find new roads still visible even with all the dust on Mars. To make it more interesting, what if we saw vehicles on the roads? Even with this information, there still are more than 10 who believe Mars in uninhabited and never had people on its surface or underground.

Lines Mean Something-What are the lines in the Martian picture following? The lines go miles and miles, traveling over the top of crater blast areas. The set of furrows on the Martian landscape pictured extend in an absolutely straight line. No naturally occurring phenomenon can account for the lines that continue over mounds and through troughs and we know that they were made in recent times because the meteor blast marks are below the furrows and none are on top of the strange lines. It almost looks like a huge farm. The deepest furrows are towards the center of the picture, but you can see many more furrows below the major ones all extending in the same direction. No one knows what they are, but they were made very recently. One reason would be as tracks of some type of equipment whose wheels make a temporary roadway in the sand.

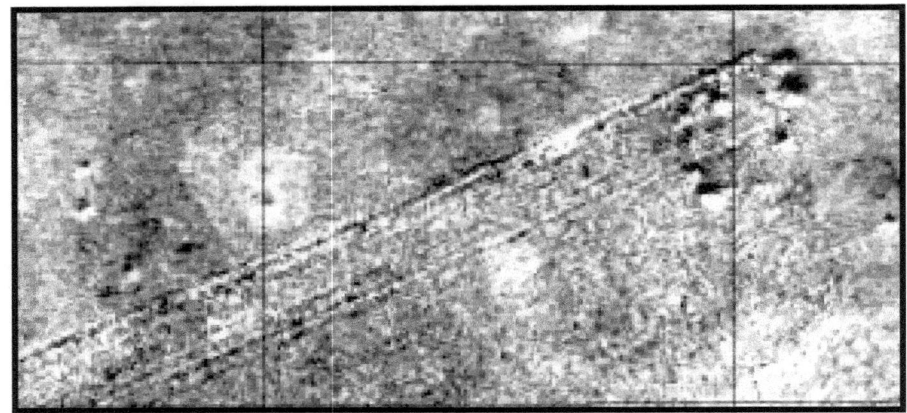

Four more roads were seen to go long distances but what travels on them was a mystery.

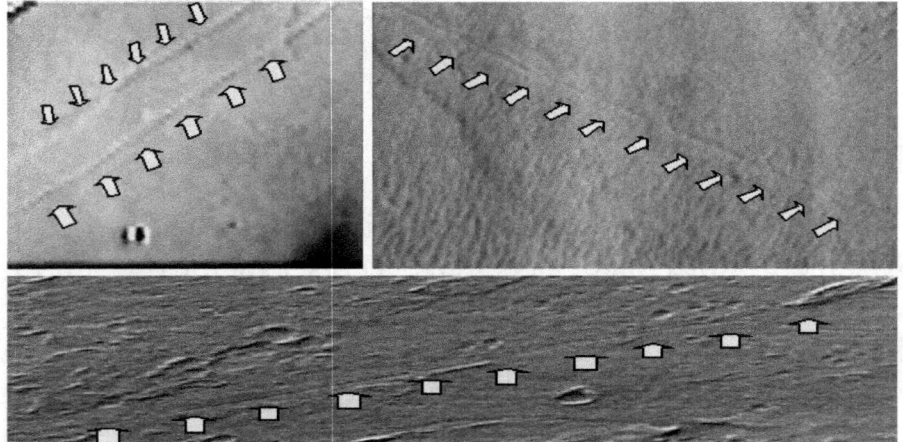

The following image shows some type of vehicle going across the Martian surface. If you look closely, you can see there are actually 3 or 4 sets of tracks so this is a well-used highway. OK! I mean well used for surface travel on Mars. Please notice how this rover and its track and the one I pointed out on Lunar look similar.

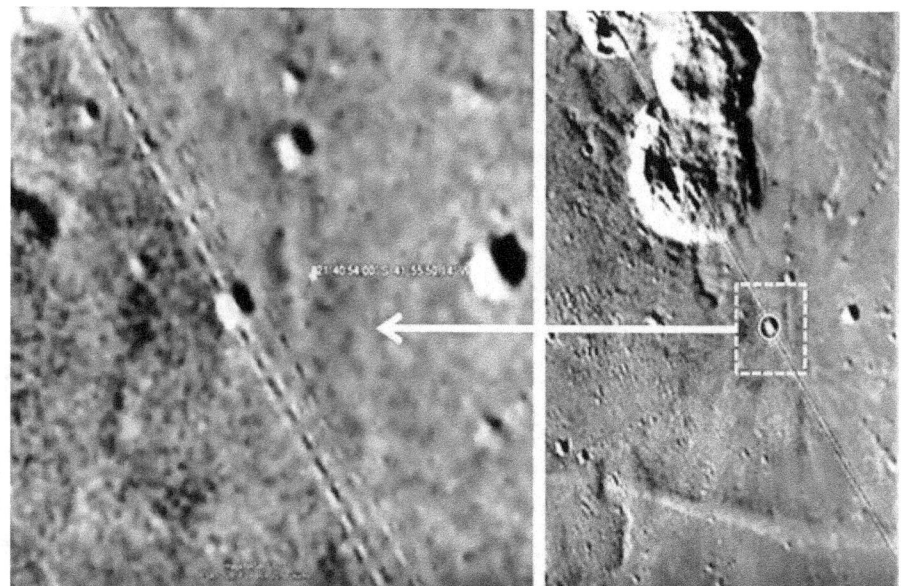

On a similar roadway, we find something decided to go a different course.

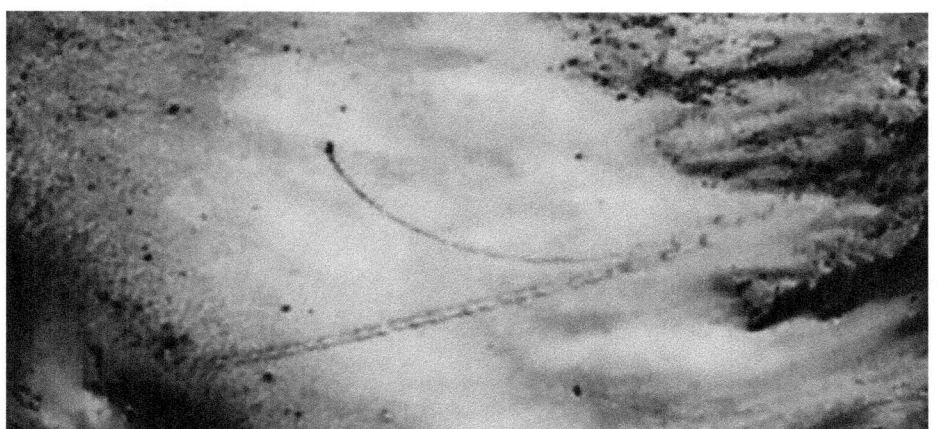

Could There be Vehicles on These Roads?

Certainly, light can play tricks on what you see from millions of miles away, but the next two images seems to look like vehicles of some kind.

 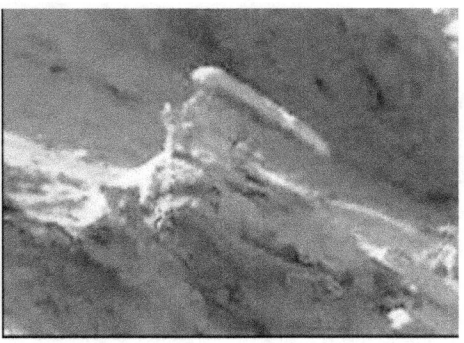

This second one even shows its "track" or what appears to be one. Don't ask me what is on top of the vehicle either. Maybe it's a weapon. Below are more of the same. I don't know what type of moving objects they are, but they don't look like objects from an uninhabited planet. I think I have seen some muscle cars that look like the last object.

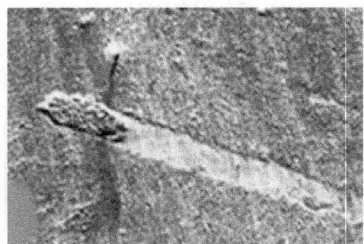 

Anomalous Artifacts

Below them are discarded artifacts including pipes, a Stonehenge, Indian burial markers, 4-balls monument, and even a discharge valve almost completely buried in the sand. If you have trouble making out the first one, it looks like a completely circular valve handle. Possibly someone used this to turn off water at one time. Now it is not needed.

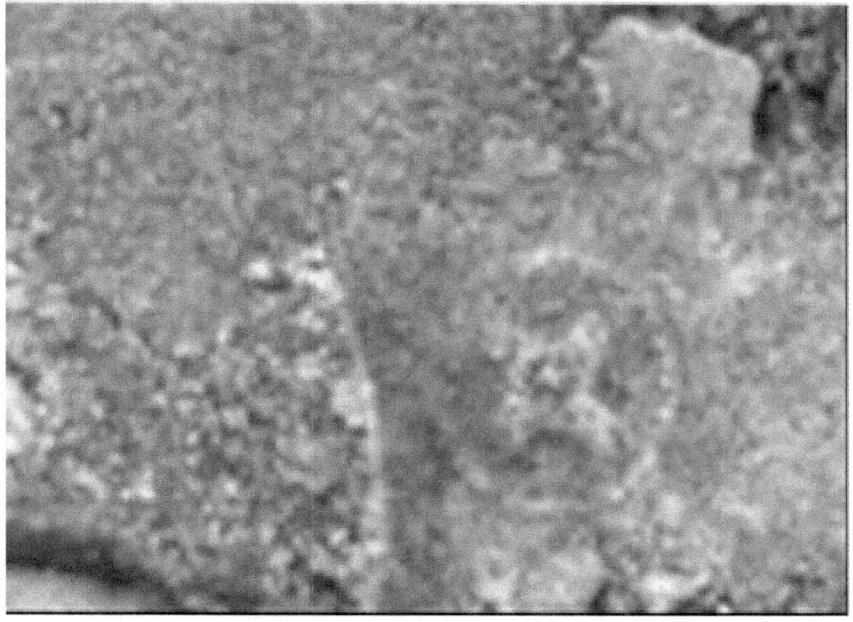

In the next group, we find all types of Martian things including their own Stonehenge. [See following right] To the left looks to me like it is some type of Indian dream-maker or some similar quasi-religious construct.

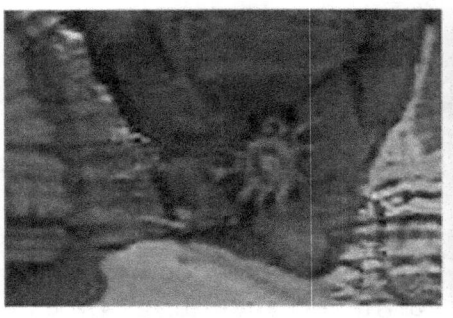 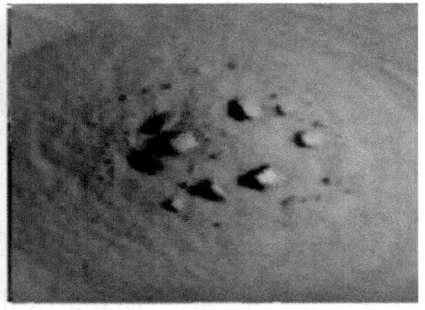

The reason I put the thing below left in the book is it appears to be metal. Possibly it is a knob off a radio or whatever, but one would not believe it is natural. The last tube thing is another mystery but it looks manmade so it's in the book.

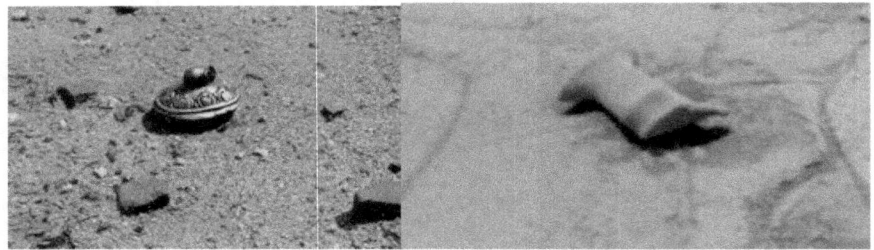

On top of a hill the rover took a picture of a dome with a house nearby and a rambling road down the hill. The dome has debris on top of it today, but possibly long ago it was a planetarium showing people where goo old earth was in the sky.

We could go on and on looking for more signs of human occupation in this section, but I think I will let you do that on your own as we look other places. I have no idea what this next thing is, but it is not naturally formed. The massive 4 balls in a row thing and the arm coming out to touch the thing along with the bridge like artifact that all this stuff is on could not be naturally occurring. The image to the right is some type of tower, but this time some massive light source is shining out over the Martian surface.

Q-TIP

The next strangeness I call the cotton swab. This massive suspended arm could not have been made accidentally by Mother Nature, but I have no idea what it was. Maybe it was some type of tuning fork or something is buried underneath that would give us an explanation.

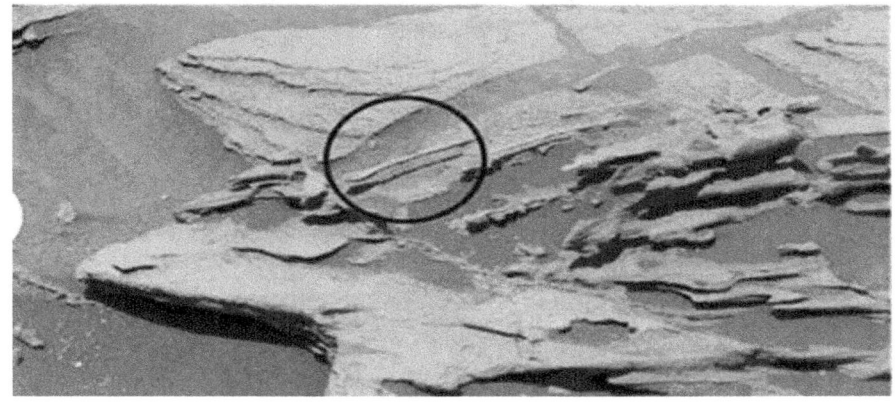

OK! Friends, just look at this stuff. Everywhere you or the rover looks, it finds more stuff. Here are some more. I call the first one "double right-angle monument". The second one is still intact pyramid. On the second row, we have remains of another vehicle of some kind and the last one is a grid work that does not look "out of this world".

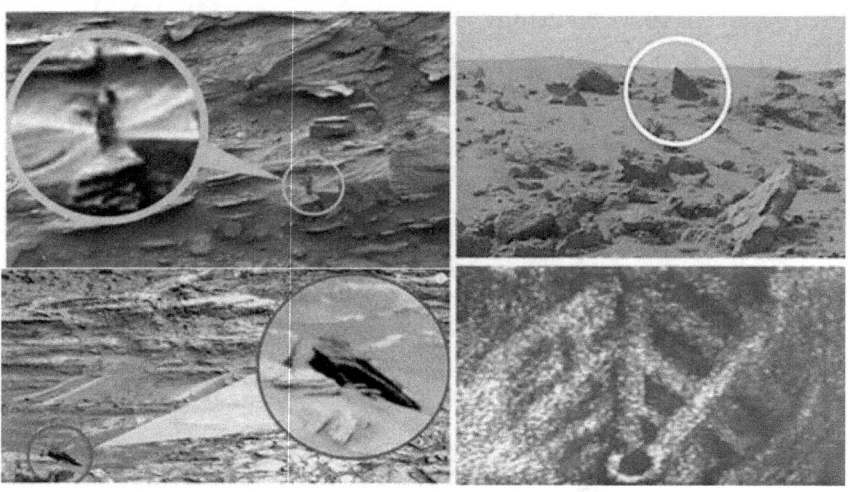

I wanted to show the next image next to the bronze army of China. Do you see any resemblance?

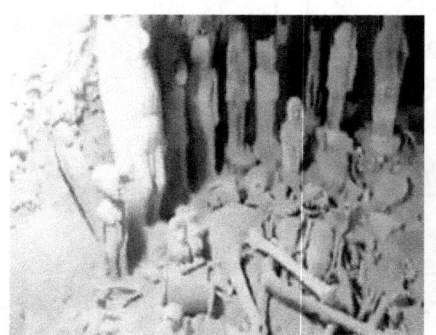

Martian Flight Anomaly

Certainly, someone would believe colonists were once on or even still on Mars if we find a flying object in the sky. What if we knew of a half dozen flying objects. Could anyone not say someone is on the planet. The answer is many still believe it is a dead planet and never was colonized.

The rover caught Martin UFOs as it went all over the place looking for things to photograph. Here is a half dozen, but there probably are more. Notice the shadow on the third flying thing below.

The first three are above and the next three are shown on the following page.

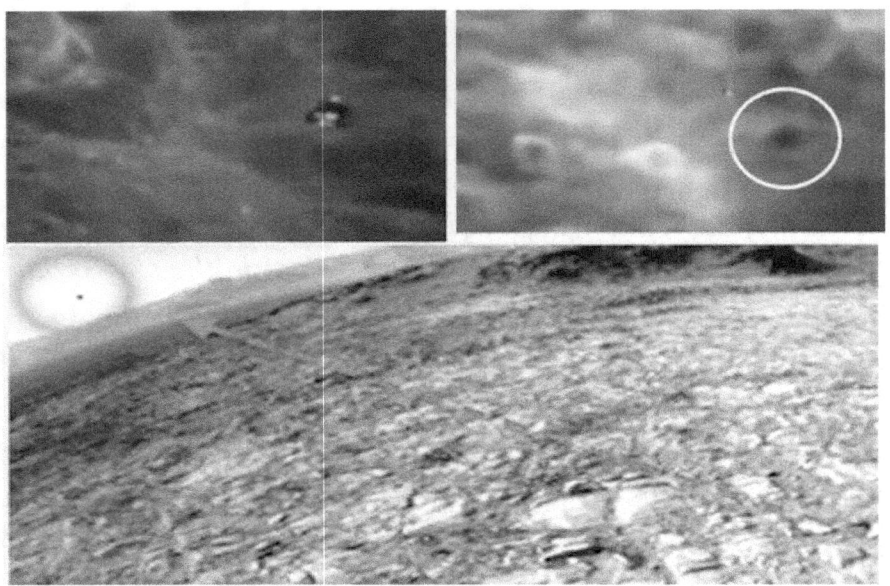

The next image seems to show some type of stabilizing Rutter. To the right is another just for good measure. Like the others whatever is flying around is seen often.

With flying machines, one would think there would not be many roads and you would be right. Possibly crazy person would occasionally have to go topside where the air is gone and he might scoot across the barren features before disappearing underground. Just like the Lunar colonization,

While Mars would have had breathable atmosphere in the early years, it is all but gone today.

Tailfin of an Aircraft? - Our little robot on Mars may have seen a piece of some type of machine lying in the desert. I don't know what it is, but it certainly does not look natural. [below left]. Other debris appears to be flying machine parts as well. Maybe they are discarded washing machines, but they look interesting.

The image to the left in the next collage seems to be a space ship, but it might just be your normal smooth, rounded surface flying saucer shaped rock so no one would have to be concerned. To the right is something hovering over the ground. Some may claim trick photography, but the rover is just wandering around.

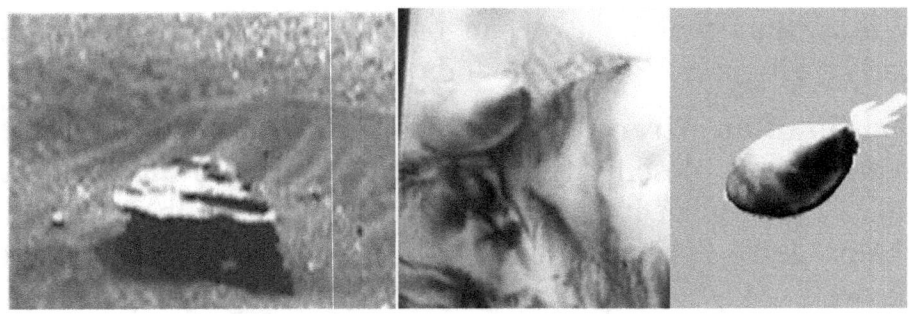

The next two objects also look like they could be flying machines similar to those we see almost every day on Earth.

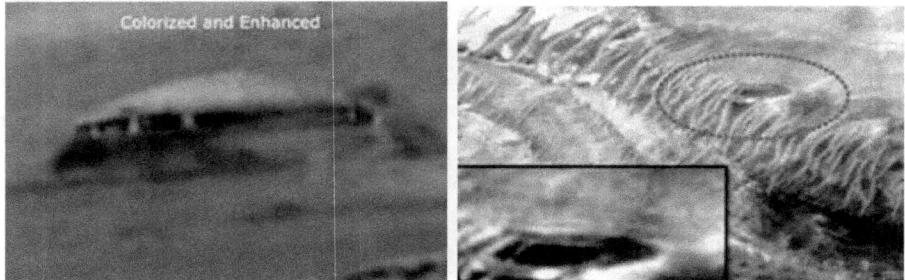

What about crash landing indications? Here are a couple.

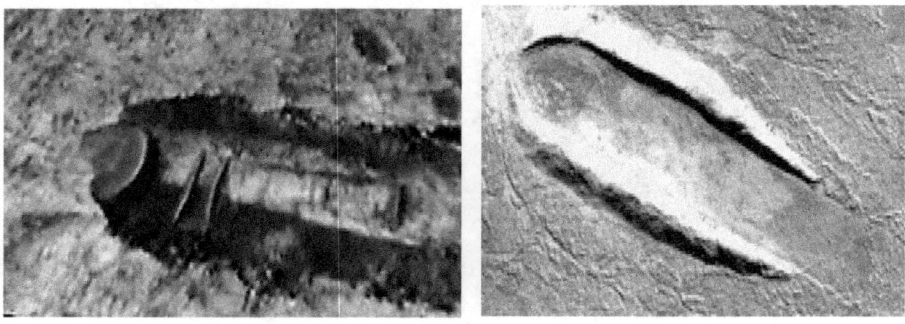

The last image I call "*the perforated box in a mountain wall where there should not be one*" image.

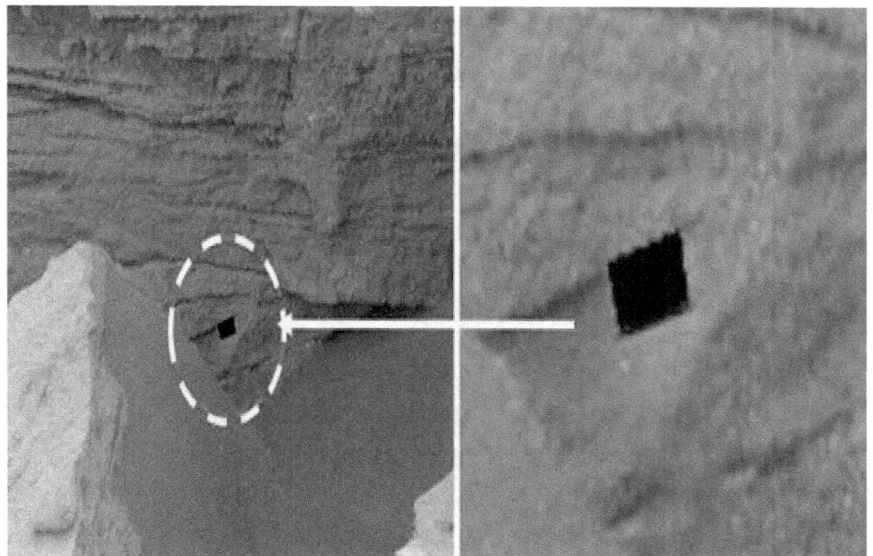

With flying machines, buildings cities, roads and the rest, someone might believe there certainly used to be life on Mars, but is there life still there?

Martian Life Anomaly

Methane on Mars

Besides all of these signs, people still have a hard time with Mars ever being inhabited, but there is methane. We find a significant amount of Methane of Mars along the entire equatorial regions as shown in the graph below. Methane is produced by animals, so animals must have been there recently. It is in significant amounts so the animals may still around. Possibly tiny micro-organisms, but it does show there is still life there.

Blue Balls- The tiny blue round things shown next were seen by the "rover". These may be colored and made to be spherical by micro-organisms that produce methane.

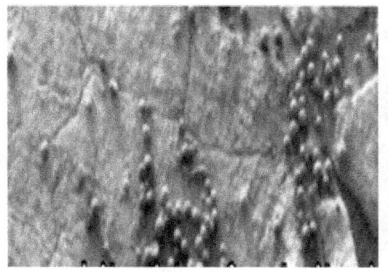 

Tread Eating Organisms-Another thing to consider is that something is eating the tracks on the rovcer. Soon it will not be able to move at all. These little buggers are hungary.

Finally, we need to look at a worm similar to that found on Venus. Notice that a few minutes after the first image, this thing came to the surface

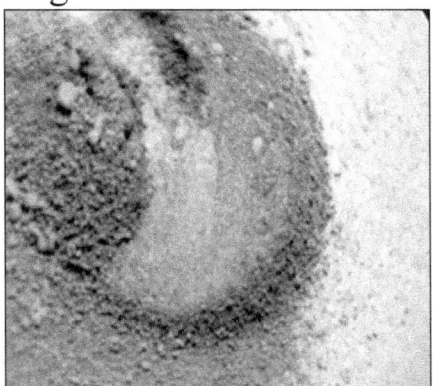 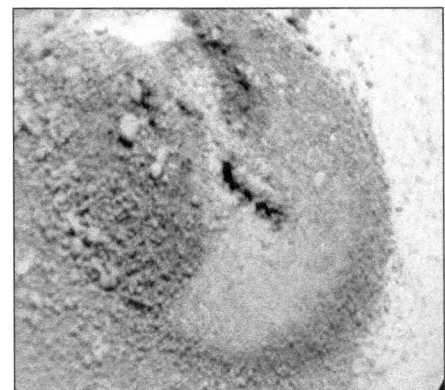

A second possible worm-thing may have come out of the ground and found on the surface as shown next.

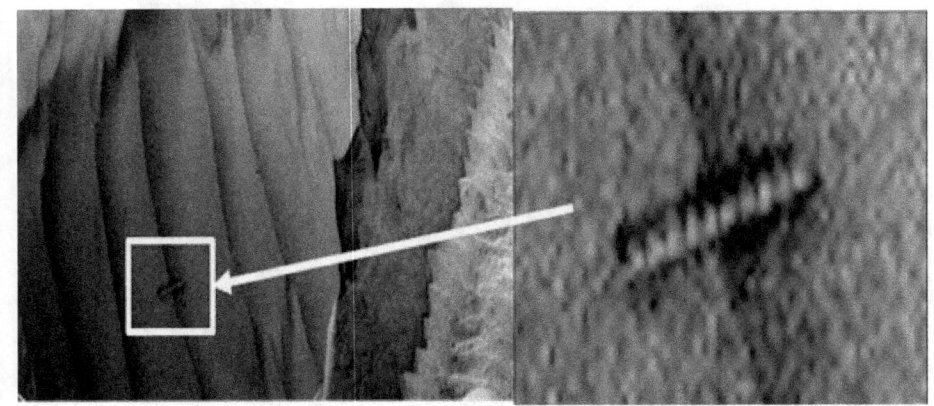

You know what they always say! If there are animals on the planet there must also be trees. As you would expect to find in a place loaded with carbon dioxide, Mars is running rampant with vegitation. Just a few of the hundreds of miles of trees and forests are shown below, Certainly our children are being told of this so they know about life there.

Martian Trees Anomaly

Certainly there must not be trees and forests. That would make it look like Mars is inhabitable. I' sorry to break the news but vegitations seems to be everywhere. The issue is that the oxygen we can believe is produced cannot be held in the thin atmosphere. Therefore a way to capture the oxygen is needed. Some might believe undewrground cities could secure the oxygen.

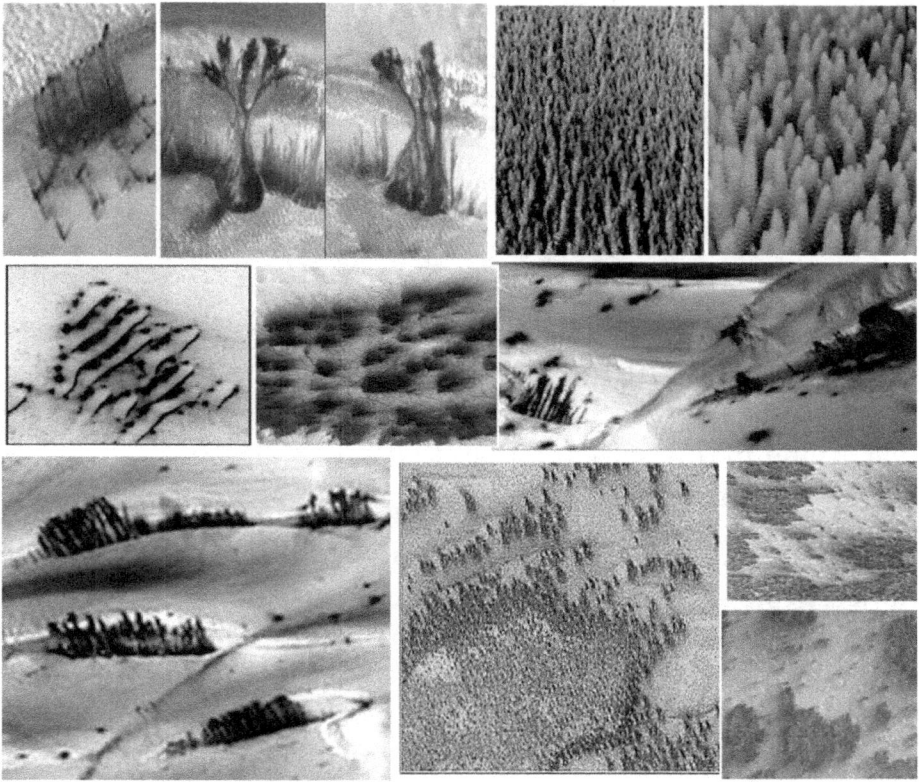

Having a reasonable timeline of some of the events that occurred on Mars suggests that it began losing its atmosphere about the same time as the earth began to spin faster during the Mesozoic Era. If that is so, humans stationed on an outpost such as Mars would have tried to conserve the atmosphere by means similar to our terraforming concepts.

Martian Terra-forming-The humans on Mars may have begun the repair of Mars in our fairly recent past or Mars may be fixing itself. It's been a long time since life could exist on Mars, but there is new evidence that life may be reemerging in the form of some type of plant life. Below is an area many believe is covered with some type of treelike plants. <u>The color and density of these things change over a yearly cycle and they seem to have limbs like trees</u>.

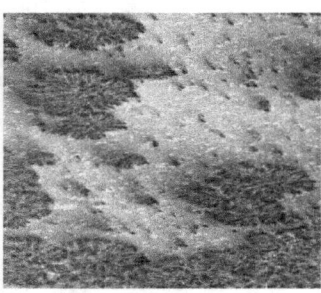

A larger section of this crazy looking area is shown on the above right. There can be little doubt that something is or was living in this area and that it is a huge area.

Dune Trees-As we look at the huge expanse of dune areas, something curious can be found at the base of some of the dunes. Round patches in the Mars photo shown above right look very much like some type of vegetation to me. The next 2 pictures are special with regards to trees. Look at the dead trees everywhere.

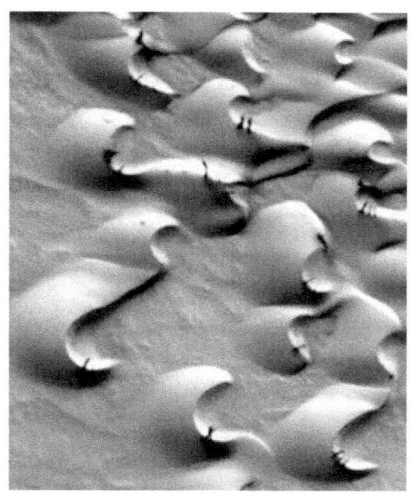 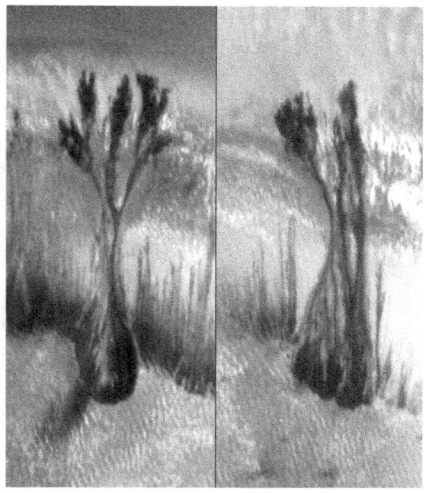

The trees in this area seem to be barely clinking to life. Somehow the tops of these dunes allowed their survival, but we could hardly call this a forest. The next picture gets us closer. If we blow up a section, the trees can really be examined easily. The trees almost look burned up. Still another section seems to show these plants form on top of small hill areas and not much life is away from the hills

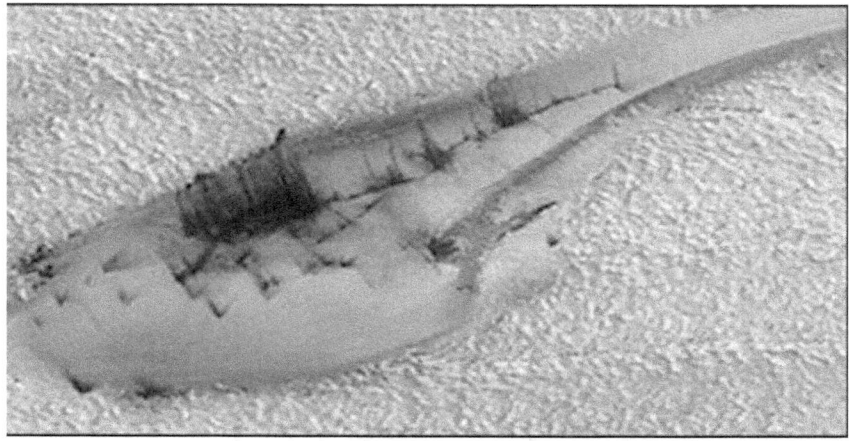

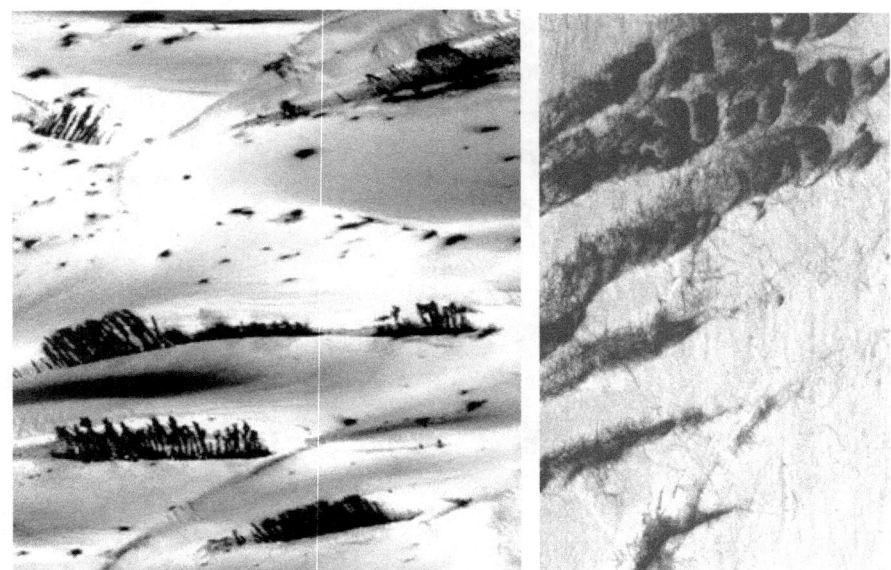

You know the old saying. If you have burned up trees, humans must have been there. While the trees look burned up, possibly that is the way tree look like on a planet with almost no air and almost no above ground CO_2 respirators. The picture [above right] may also be dead trees of sage brush or something similar. We see the crisscross hairy strands of this strange, plantlike array. Possibly it shows strings of mineral deposits that form these strands, but to me, they look connective and bush like.

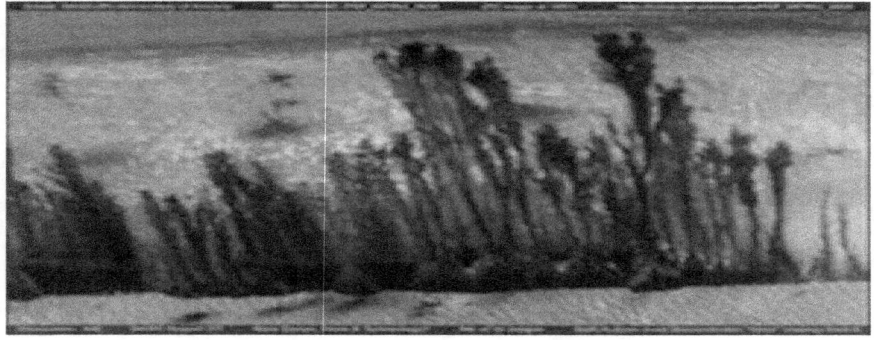

Martian War Anomaly

Look at some more of the curious details found by the many researchers in this area. The top row of the following collage shows melted bricks. Can there be any reasonable doubt that these structures are regularly spaced, rectangular, and have high walls just like you would expect in a city. Even what appear to be sectored roadways can be made out. I know it looks all melted, but that raises an even more difficult question. What melted whatever we are looking at? I believe that the melting was intentional. As the average temperature on Mars is very low, a reasonable belief might be that some type of nuclear discharge would have been used to melt all the stone walls---especially when Xenon-129 is found.

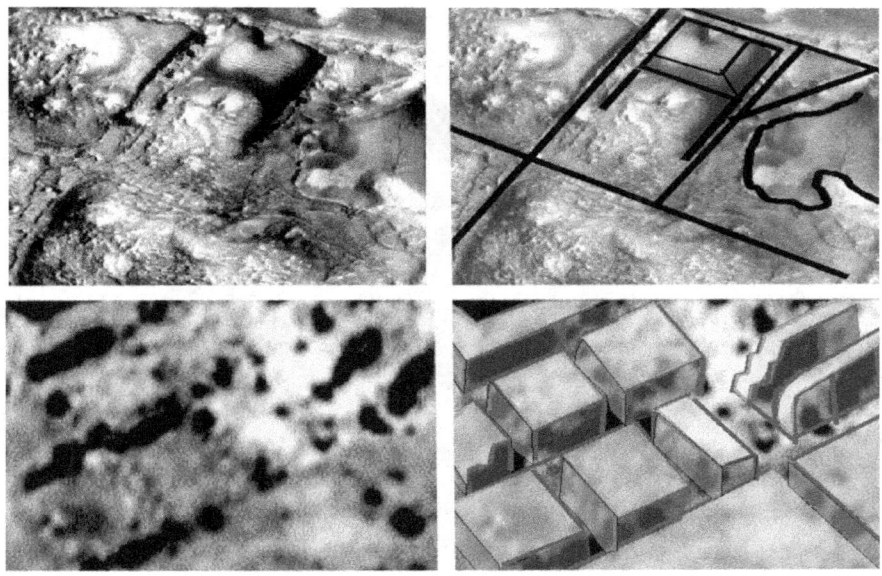

The preceding images are from 2 completely separate areas on Mars so the melting continued. It looks almost like the cites had been in a war. On the images below we find the remains completely gone except for the base that is almost completely covered. The bottom image has a similar scene of destruction. These show the remains of war being covered over by sand and dust, but the weapons left something behind.

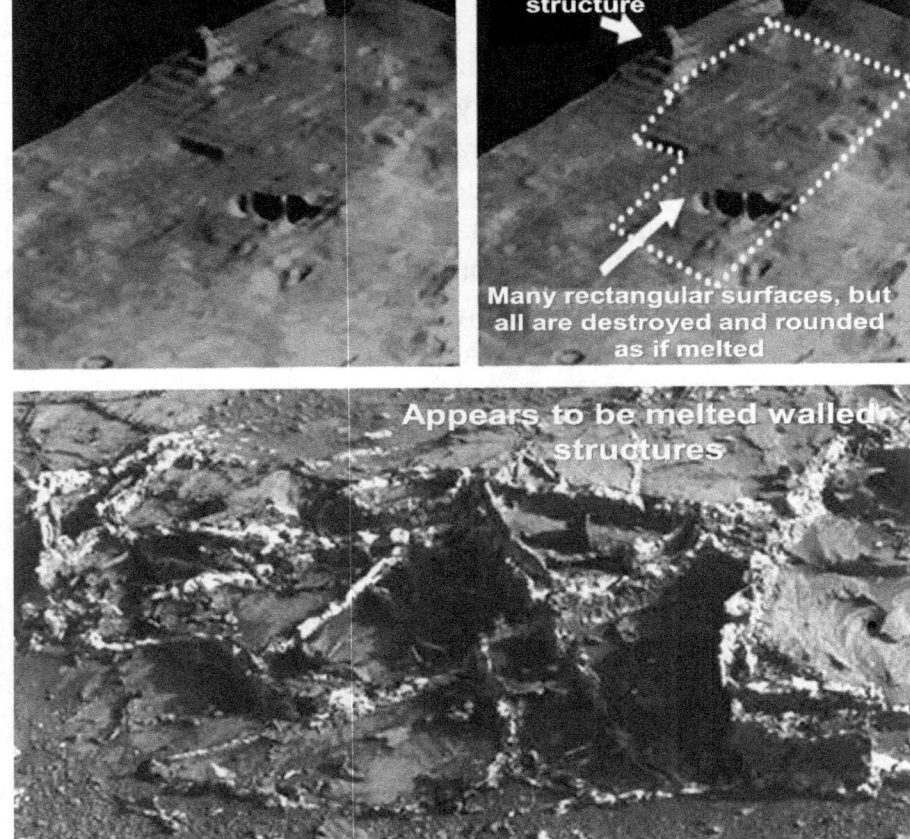

Another couple of areas that appear to be completely destroyed should also be recognized. The following set of six square things was probably made by Nature, but that would ONLY have been if Nature was some Martian colonist's name. Good old "Nature" loved to make weird things. Possibly this is the remains of some ancient port, but not much is left as it has been almost completely destroyed. To the right is a square thing almost completely covered by the sands of time. While it is thousands of years old, there is no question a huge square structure once was on the Martian surface at this location. As the walls are ground down to almost nothing, we can believe the building complex was destroyed during some horrible war.

 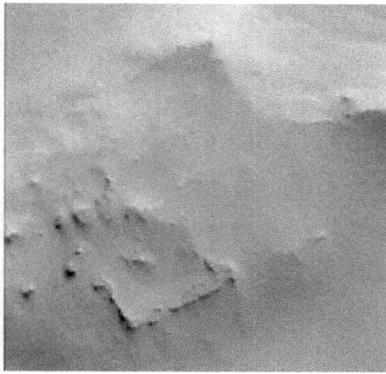

Xenon-129 Evidence

The evidence strongly suggests that many of these Martian colonists left their homes in the not too distant past. This Xenon-129 stuff is a "second order nuclear fission by-product" and guess where too much has been found. Mars is the winner. It has this nuclear by-product in abundance. The problem is that no one can determine how nuclear blasts went off on Mars. Don't believe stories of Mars collecting the material from some ancient supernova because it just doesn't make sense. What does make sense is nuclear bombs, but then there would have to be people there and someone must have been mad at those people. Some of the scientists trying to figure out why this Xenon 129 is there don't believe in the very recent nuclear explosions on Mars, but they are completely baffled as to what else might have caused it. That type of stubbornness keeps our history books comfortable and useless.

What Does All This Mean? - Let me tell you a short story. The ancient people of earth knew of the upcoming destructions associated with the Cretaceous Extinction and had 2 choices. Stay and probably die or go to the Moon, Venus, and Mars to start new colonies. During the Pleistocene Extinction, there would have been just as much incentive to colonize. Hopefully, I have presented enough evidence to show you they could do this and certainly there would be incentive. Besides the normal Flying machines, they probably had a better handle on teleportation than we have today. While recent experiments in the Canary Islands and Korea both have shown instantaneous transport to another location and we soon will be able to use this for a number of requirements. My personal belief is that during ancient times they had gone farther in those experiments. If I were a betting man I would say that the really large Anak or Annunaki people went to Venus that had the oxygen levels needed for their large bodies, while Mars would have colonies of mostly normal sized people that could survive with less oxygen and more easily establish underground living spaces. We can believe the air was breathable on both planets even up to the Pleistocene Extinction. No matter how and when, the breathable air is, now, all but gone. Many would have left a long time before the air became too thin. If some stayed, they would have had to stay underground. Perhaps they would have to make sealed tunnels that were somewhat clear to let in the warming rays of the sun. If we could find these tunnels, we might even believe, life may still be on Mars.

Tunnel Anomalies

If you say it didn't matter that there was water, its simply too cold to live there. The answer might again be just below the surface. What if they typically lived underground and had surface tunnels to allow comfortable transport. OK! Not completely comfortable in the frigid lands. If we look at the full length, we see crisscross avenues and entry into deep caverns. Possible protection from a planetary atmosphere that was getting thinner every year. Notice how several of the tubes are going down away from the surface and into the planet interior. Possibly the tubes also were solar energy solctors and needed to have surface view. Maybe, keeping people below the surface for long periods of time was not a good thing so surface areas were designed.

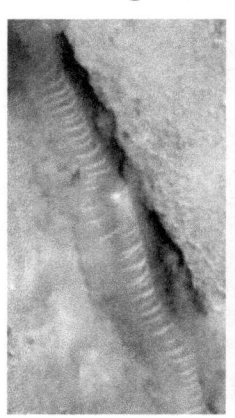  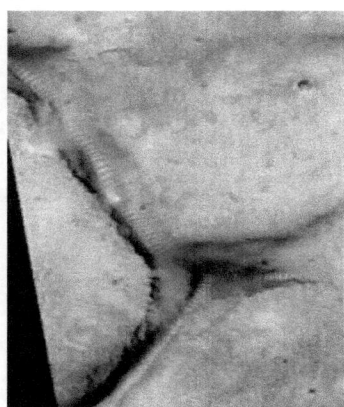

Still another section of protective tubes has also been uncovered over time. Where they go, noone knows, but they seem to go somewhere. If we look at the sections of the preceding tunnels, we see some sections that are constructed in an accordion shape to allow for substantial expansion.

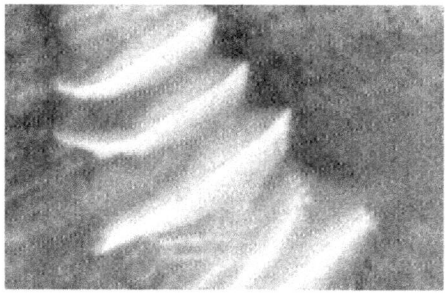 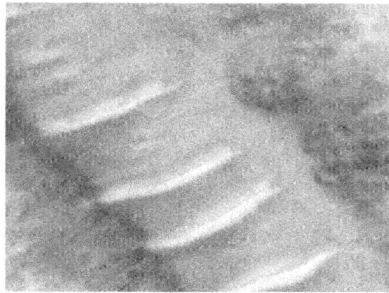

Other sections reveal heavy duty rib designs to insure maximum structural integrity.

Some Failures-Let's look at the artifact next which is part of the MOC image M1501228. Many believe this to be a huge tunnel that is about 300 meters high and 200 meters wide. It is ribbed for strength like the others, but this one didn't hold up. It was breached a long time ago. Maybe the colonists on Mars were living underground during the initial war years just like the people on earth. With radioactive xenon and melted cities, the tunnels and underground living might have been their only hope. Reflecting light off the surface of the tunnel is evident in the photo which suggests a very smooth surface. No one knows what the tunnel was used for, but not many dispute that this feature would be almost impossible to be produced from natural phenomenon. If this rupture had happened hundreds of thousands of years ago, the cavity would have filled in and there is clearly an open area below the covering, so the tunnel would have been operational during the Pleistocene and even into the Holocene Ages. Whatever happened to this tunnel was tremendous, just imagine the explosion as a couple thousand meters of structure gave way in a devastaing way. Certainly some lost their lives when this important Martian protection was destroyed.

Let's look at some of the details of this marvelous find. Note the clear domed passageway as it can be seen in the fissure.

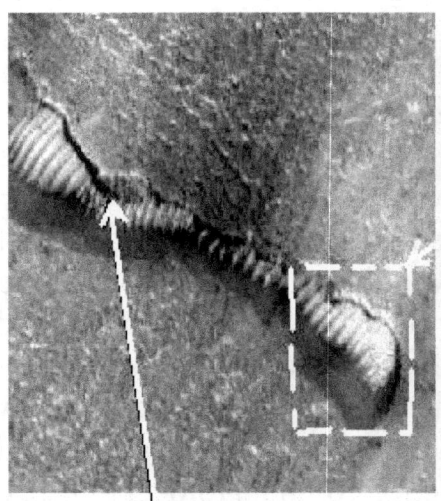

Hallow tube- well defined dark shadow inside ribbed tube

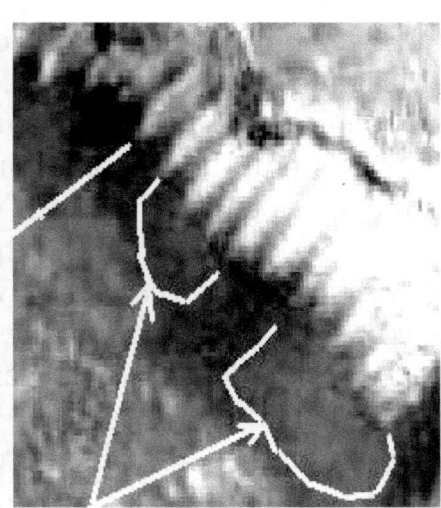

Reflections into the shadow

More pictures are below showing how many of these covered tunnels have been found so far.

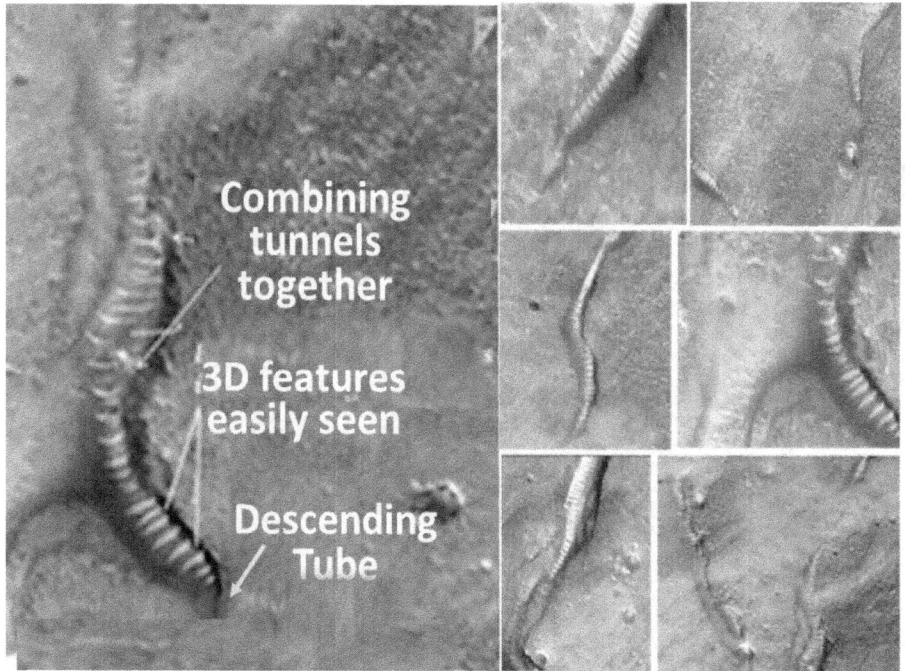

Where is the Air? - If you have trees, there will be oxygen, If the oxygen is simply discharged into the air, it will escape as the planet is spinning too fast to hold it. Therefore, we might believe someone could live underground and capture the air along with hundreds of thousands of gallons of water known to be locked in the ground a supplying needed life to the trees. We can even know where the inhabitants might still be as underground tunnels occasionally rise to the surface and sometimes they are damaged as shown in 2 of the images. The almost transparent walls allow for what little bit of sun there is to enter the tunnels to help support life and NASA has published thousands of images that keep saying the same thing. One question might be, "Are there things we are not being told?"

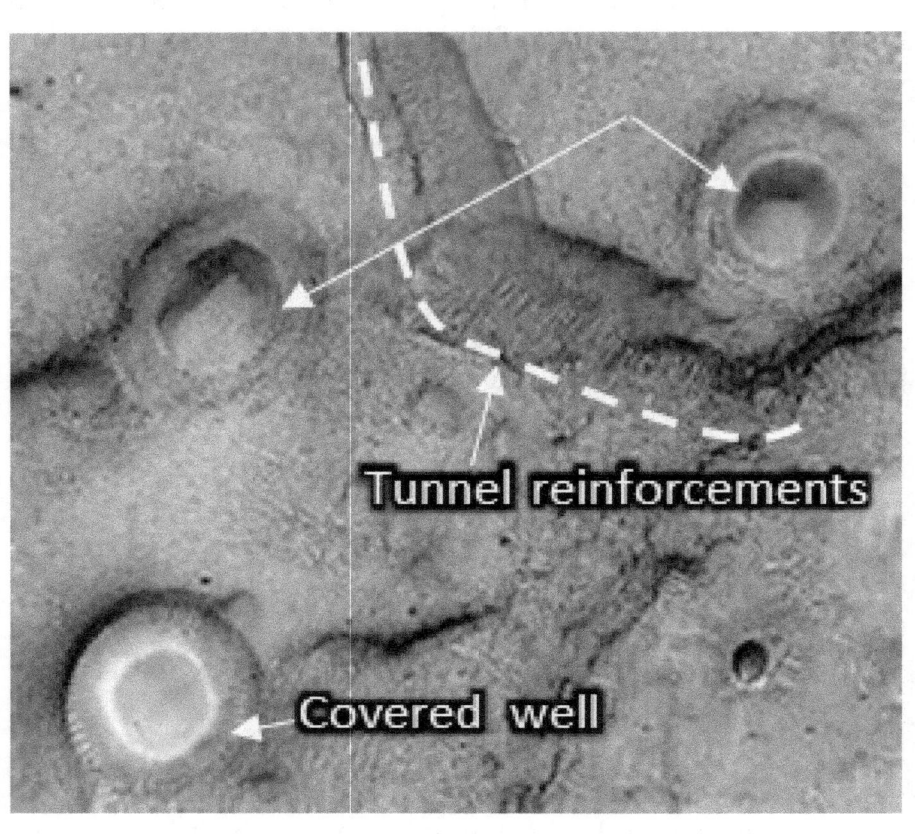

Are People There Now?

This next bizarre section would certainly be the biggest anomaly by far, but what if it has a level of truth? I certainly will not and cannot say definitively there are people on Mars today, but there are stories that we need to, at least, consider. This is a story about something called the Pegasus Project. For this story, we must first see what the Web-Bot determined. WebBot is a program that senses probabilities by information scurrying around on the internet. Cliff High is the genius behind this thing and it has been predicting all types of things, mostly in the financial world, but also catastrophic events and other things that would not seem possible to detect by listening to random words and phrases. Anyway! A September 15, 2009 report predicted that a "planetary whistleblower" would emerge from the current period of U.S. financial collapse. It also, somehow, came up with names--- Mr. Basiago, a lawyer from Washington State and Dr. Anderson a Physicist. ---So, what!!!!

Sure enough, there currently is a so-called whistle blower named Andrew Basiago that may tell us a little more about something that sounds bizarre that deals with an astonishing method for travel to Mars. According to testimony, Andrew Basiago, had provided evidence "of a sort" that secret U.S. time travel technologies were used as early as the 1960s. He was recruited when he was a teen to participate in a program called the Pegasus Project that was carried out under a US Defense Advanced Research Projects Agency [DARPA]. Mr. Basiago described probes to Mars he took via teleportation and something he called chronovision during the early days of time-space exploration by the US government between 1969 to 1972. This Pegasus Project was somehow associated with the Ari Force, DARPA, and the CIA and it may still be operating. Of course, the current Quantum Mechanic Time space jumps confirm the possibility, but the only things we hear about are teleportation of sub-particles. In this project, there were teleporting entire people and transporting them to Mars or even back in time.

I know all this seems bizarre and wacko, but there is more.

Like I said, I know everything here sounds hokey, but there are many other whistleblowers.

Dr. Anderson-His name is Dr. David Lewis Anderson, director of the Anderson Institute. He actually came out before Mr. Basiago and provided a two-hour interview on December 23, 2009 to give an extensive account of his time control research for the U.S. Air Force, which he later continued at his Time Travel Research Institute and other organizations. He is shown middle above. According to this physicist, he was employed at a young age by the U.S. Air Force conducting advanced research and development at the Air Force Flight Test Center at Edwards Air Force Base in the Mojave Desert. During that time, he laid the foundations for what would be known as "time-warp field theory," an approach that was actually used for time travel according to his testimony.

Mr. Cooper-Naval Intelligence officer Milton William Cooper corroborated U.S. Mars bases and time travel accounts of DARPA whistleblower Basiago, Dr. Anderson, and CIA whistleblowers Bernard Mendez and William Stillings. He also admitted to key aspects of accounts made by DARPA whistleblower Andrew D. Basiago and the CIA whistleblowers regarding U.S. secret bases on Mars and secret U.S. government time travel capabilities. Milton. Cooper stated that the U.S had first landed on Mars on May 22, 1962 and that, by the time the U.S./NASA public space program landed on the moon in 1969, the U.S. already had a moon base there, since the mid-1950s. Here are some of his own words. *"The first moon landing was May 22, 1962 ... or excuse me, that was the first landing on Mars. I'm sorry, May 22, 1962, was the winged probe that used a hydrozine propeller, flew around approximately three orbits and landed on May 22, 1962, was a joint United States/Russian endeavor. The first time that we landed on the moon was sometime during the ... probably middle 50s, because at the time when President Kennedy stated that he wanted a man to set foot on the moon by the end of the decade we already had a base there."*

Mr. Relfe-Michael Relfe is still another "informer". [See preceding right] A former member of the U.S. armed forces, he was recruited to go to Mars in 1976. He stayed there on a secret base for 20 years before returning by one of those teleporting time traveling things, but here is the weird part. In 1996, he supposedly was sent back to 1976 to finish his military tour. While it may never be known the validity of Mr. Basiago's, Dr. Anderson's, Milton Cooper, Bernard

Mendez, William Stillings, or Michael. Relfe's statements, one can be pretty sure that the time-travel" Rainbow Project did not die as the USS Eldridge disappeared during World War II. [We remember if as the Philadelphia Project.] Maybe there is more evidence that can be validated better.

Arthur Neumann and Project Camelot-Seems like when one whistleblower comes out the field becomes ablaze with those no longer afraid. As reported on "Nolies Radio", another witness indicated that there is a colony on Mars. His name is Arthur Neumann. [See preceding middle] As the whistleblower who reportedly worked in a program called Project Camelot stated on July 25, 2009, at the European Exopolitics Congress in Spain; there is life on Mars and there are bases on Mars. He went on to say, *"I have been there."* He also provided details of teleporting, a permanent Mars colony, and participating in a one-hour project meeting, which was also attended by representatives of an intelligent civilization that lives in cities under the surface of Mars. This was during a July 26, 2009 Futuretalk interview. As part of his job, Mr. Neumann spoke about the details of his teleportation to Mars during a one-hour project meeting attended by what he called "Martian humanoids" at the secret underground Mars colony. This not only adds to the growing number of eyewitnesses to some type of teleportation and time travel, but it also brings us to the great granddaughter of President Eisenhower.

Laura Eisenhower- Laura Magdalene Eisenhower [See preceding right] indicated that she was a survivor of an attempted recruitment into the secret Mars colony between 2006 and 2007. All 7 of these witnesses seemed to be talking about the same thing. Some group had developed a way to transport through time and space. I know experiments in Korea and on the Canary Islands have proven that space time can be eliminated in transport of subatomics, but this is with people. Pretty impressive.

> *I don't think I could describe anything that sounds more absurd. All I can say is the more you hear seemingly ridiculous things and those things do not contradict other statements, the more the things could be true.*

Mr. Basiago stated that in the early 1980's, when they went, the U.S. facilities on Mars were rudimentary and resembled the construction phase of a rural mining project. While there was some infrastructure supporting the jump rooms on Mars, there were no base-like buildings like the U.S. base on Mars first revealed publicly by Command Sgt. Major Robert Dean at the European Exopolitics Summit in Barcelona, Spain in 2009. On Mars, they encountered primitive conditions and some type of dangerous animal that evidently killed a large number of the participants. In class, they were told that *"Of the 97,000 individuals that we have thus far sent to Mars, only 7,000 have survived there after five years."* While it may never be known the validity of Mr. Basiago's, Dr. Anderson's, Mr. Relfe's, or the other whistleblower statements, one can be pretty sure that when the USS Eldridge disappeared during World War II in Philadelphia, it was not the last teleporting experiment done. It seems the Pegasus Project took over where it ended. For a moment, let's just say that these descriptions have a level of truth in them.

Anomalies Beyond Mars

If people colonized Venus, Lunar, and Mars, what about the other planets. In general, the answer is the others are not fit for humans, but there are some exceptions on moons that have water and sometimes even oxygen. What we have found on some of these seems to tell use that people not only flew by these planetoids in search of living possibilities, that there may have even been fighting over some of these areas. One of Jupiter's moons, named Ganymede and one of Saturn's moon's, named Titan, hold claim to being larger even than the planet Mercury. These larger moons around the monster planets are big enough to have an atmosphere. If humans found water, they may have looked for life or even visited the frigid planetoids. Some of the terrestrial planetoids are moons of Jupiter, Saturn, and Neptune. By far, Jupiter has the largest quantity of large planetoids around it with four, but let's start with Neptune whose major moon is named Triton, not to be confused with Titan that circles Saturn.

Neptune's Terrestrial Moon Anomaly

Neptune is one of the massive ringed planets, but Triton, Neptune's moon, is interesting and we find a little bit of a surprise there. It's not formless as one might think nor is it just a chunk of rock. It has extensive ridges and valleys and appears to be inhabitable as the planetoid is 25% water and is classed as terrestrial. It also has a <u>mostly nitrogen atmosphere</u> and <u>has organic compounds existing on its surface</u>. It has few craters and the surface has a changeable Polar Ice Cap. So here we have seasons, organic compounds and an atmosphere that is nitrogen based just like earth. All we need is heat that comes in the form of volcanoes found on its surface. One slight disadvantage is that the volcanoes expel water rather than molten material. We can say humans may have stopped here taken samples and left because it is simply too cold to hold habitation.

Saturn's Terrestrial Moon Titan

At first Titan seems like an ideal planetoid for habitation. It has some of the requirements of life and the really neat thing about this place is that it has a huge continent surrounded by liquid. One strange thing found by the Cassini satellite is a wavy bluish atmosphere. [See below left] Once we got closer we could see many lakes all over the place. Titan contains water and is therefore considered terrestrial; however, it is not likely that it contained life. So, you might ask, "Why?" It all has to do with the oceans on the planetoid. Dark regions on the surface indicate massive oceans surround large continental masses. The problem is that these oceans don't contain fish because the <u>oceans are made of Ethan</u>. The Atmosphere is <u>mostly Nitrogen like ours</u> and organic compounds are in reasonable supply, so <u>simple organisms possibly exist</u> so long as they stayed away from the oceans so we might suggest humans did land here and take samples, but they stayed away from the oceans and left quickly.

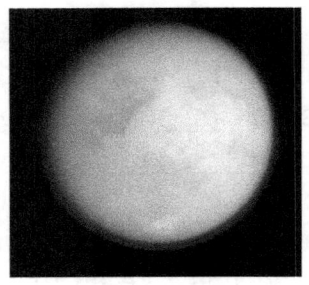 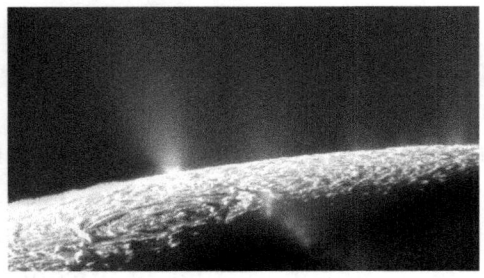

Saturn's Terrestrial Moon Enceladus

On this strange planetoid there are linear sets of grooves tens of kilometers long that traverse the surface. The uncratered regions are geologically young and may indicate that Enceladus has experienced a period of relatively recent internal melting. While it is the smallest of the 6 major moons of Saturn, it <u>is apparently heated by reactions with its closely linked cousin Dione</u>. The satellite is about 500 km in diameter and is shiny as shown in the preceding left image. When I say shiny I mean it. It is 90% the brightness of a mirrored surface, so it is surfaced with a beautiful layer of snow and ice. If we look at a close up of the planetoid, we find that around some of the volcanic craters are refrozen areas that were once liquid water. It <u>may still have liquid water near its surface</u>, which makes it a good candidate for supporting life. When a planetoid has enough heat to do that, it can sustain life. No other signs have been found, but it is way out there and it's hard to get reasonable detail. One thing strange is massive is volcanoes spewing into the atmosphere as shown on the preceding right image. Possibly humans landed on this planetoid took samples but they would not have stayed long.

Jupiter's Terrestrial Moon IO

Not only does this moon have a weird name, Io looks weird. It has a sulfur dioxide atmosphere and not much water. Some don't consider it to be terrestrial, because the surface is covered with silicate lava flows from the huge erupting volcanoes. [See following left] I brought it up because it is huge and it has something special. Io is warm enough for life and its complex chemistry make it suited for establishing some forms of life. Humans could have stayed here for short periods of time, I suppose.

Jupiter's Terrestrial Moon Calisto

Although this possible outpost has water, it is all frozen as you would expect. It is almost completely covered in ice. There seems to be no volcanic activity, the atmosphere <u>is mostly carbon dioxide</u>, and there are a fair number of craters around the entire surface from meteor strikes. It doesn't appear to be inhabitable [See three images below], but there is something curious that was found.

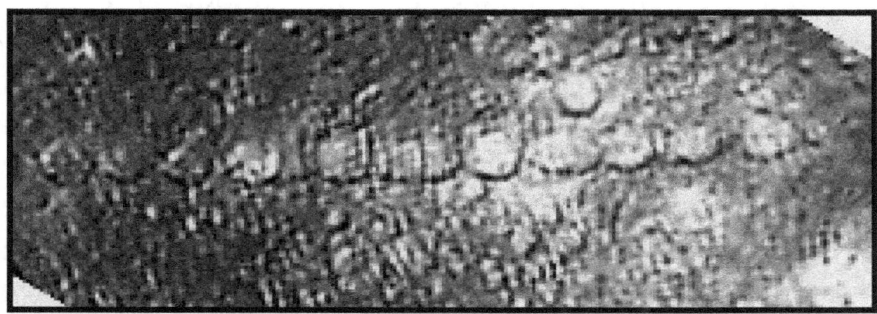

Blast Craters may show attempts to halt colonization. Please look at the image above and notice that the craters are in a row, they get larger then smaller as if the entity depositing these holes was strafing the planet. Like images on Venus and Lunar, these appear to be blast marks from massive bombs. Why would someone blast an empty planetoid? Some try to indicate that a long single file string of meteors hit or one meteor hit and then bounced 12 times, getting heavier in the middle. Evidence suggests limited colonization and attempts at destruction by humans.

Jupiter's Terrestrial Moon Ganymede

Ganymede is interesting in that there is a huge continent on its surface which could allow for life and there are strange linear grooves on the surface that appear to be to have been place in a pattern like a farm [See image to the left] I don't know what natural event would have made the right angles and line straight furrows. Do You?

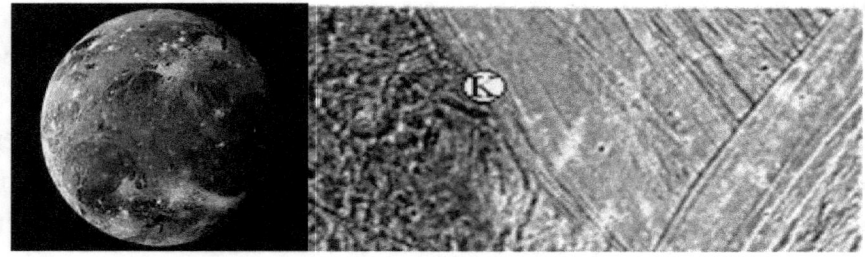

Although Ganymede is cold and has minor amounts of oxygen just like Calisto and Europa, its topography is unusually structured and the planetoid has another similarity with Calisto. We have found and photographed a long line of blast craters on Ganymede. The strike path doesn't come from a meteor that split up just above the ground in even pieces, nor did a meteor go bouncing around 16 times. What do you think these evenly spaced almost identically sized blast areas are? It should also be noted that this feature is not very old as all of the other features in the area are under the 16 blasts. It also should be noticed that the center crater is larger than the entry and exit ones like we found on Calisto. Another thing that is similar to the artifact found on Calisto is the fact that the center blast is stronger than those before or after the mid-point. Does that sound like the breakup of an asteroid or humans attacking other humans that might have been on the surface? [See the following image. Possibly some type of road is crossing the path of the blasts.]

Jupiter's Terrestrial Moon Planet Europa

Europa is the smallest of the 4 major planetoids of Jupiter and perhaps this place sustained life at one time. One thing that helps Europa is that it is closely united with Io. The forces of attraction between these two planetoids may cause heat and you really would like some heat when you are stationed on a planetoid around Jupiter. What we find is that lava is coming out of active volcanoes to heat the place. The surface looks like long fingers extending in every direction. These travel around the entire surface, but no one really knows what the streaks mean nor does it mean there was life. [See the image following left] In addition to the strange streaks, the surface is full of things to investigate, that look like roads that show up as parallel lines that stay the same size for miles and even tight clusters that appear to be communities. [See following right]

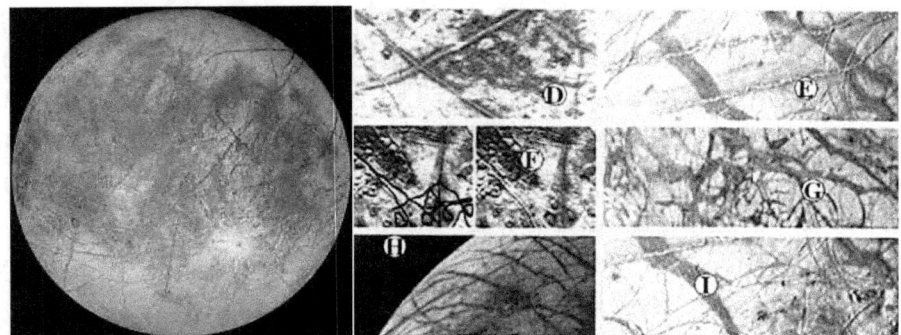

Volcanoes could have been a heat generator that aided human survival in the past. If I were going to fly to some planetoid around Jupiter, Europa would be more reasonable to me especially when we found real uncombined oxygen in the atmosphere for breathing and liquid water. Don't get me wrong, there is no certainty that any of the Jupiterian planetoids ever had life for any extended time, but, with the building blocks of both water and oxygen, they do appear to have had some human interaction in the past.

Conclusions and Discussions

First let me apologize for the section on current life on Mars by modern humans. That being said, I think we need to keep open minds concerning our historical references and not simply describe a whitewashed timeline that takes out all the meat. It is the meat of history that can prepare us. The other stuff is simply window dressing. Hopefully, you have a better appreciation for anomalous articles and less of an admiration for what we call Historical textbooks. Hopefully you at least partially believe or understand there are possibilities addressed in this book that are not anomalous and should not be pushed under a rug somewhere.

- While it is well known atomic decay timing is horribly unreliable, historians use this method all the time and being TRUTH.
- While it is well known that all the crustal material is missing from the Pacific Ocean, a crazy tectonics movement lie was developed to explain the Pacific Rim Mountains, the Ocean hole, and all the rest so a Martian flyby does not have to be investigated.
- While it is well known giant civilized people lived with dinosaurs, it is continuously ignored, downplayed, and identified as anomaly so a stupid but nice sounding uncontrolled evolutionary development of humans could be described as the only truth.
- While it is well known that hundreds of ancient texts, models, etchings, and drawings, described the ability to

flying and they would explain hundreds of anomalous historical details, no details have been put in modern, almost worthless histories.
- While there are a multitude of Biblical details about the destruction of Venus/ Rahab, religious leaders seem to push interpretations that make no sense to keep from saying Venus.
- While over ½ million craters are all tightly placed on the East Coast of the United States 10 to 12 thousand years, no history describes the horrible meteor shower even when it is described in almost all ancient histories around the world.
- While Neanderthal and Cro-Magnon Brains were determined to be larger than modern man's brain, we hear that Neanderthal and Cro-Magnon lived a Stone Age existence not even caring how stupid it sounded.
- The multiple descriptions of space travel described in ancient histories around the world are all eliminated from history textbooks to keep students from being confused with facts.
- Thousands of sightings of UFOs have been recorded over thousands of years and somehow ALL of them are missing from history textbooks.
- While is it well known that CO_2 cannot be absorbed into the atmosphere, the current Earth average temperature is lower than it was at the turn of the century, and our heating and cooling cycles align with the sunspot activity of the sun, a cockamamie "automobile gasoline is destroying our planet" scenario has been put into textbooks as fact instead of fantasy.
- While the details on the moon describe a fairly extensive colonization effort. It has been erased in our textbooks as

fantasy and the hundreds of bits of details are completely ignored.
- Flying objects, earth moving devices, flashing lights, and other detail of current life on the moon are simply ignored.
- Details of the moon's hollow areas that could provide dwelling places holding oxygen on the moon are ignored.
- The roads, noble gas isotopes, well-defined lava beds, and distinct images of rivers on Venus are never discussed even though they prove Venus was destroyed not long ago and some super ancient greenhouse event is substituted.
- Houses, and melted cities all shown possibility of colonization on Venus, but no one talks about it.
- The plasma ribbon that is still extant between Venus and Earth clearly shows the two planets came too close to one another to spark the planet's demise, but that would be an unthinkable thing to put in a TRUE history.
- The details of the industrialized Martian planet, homes, and ancient towns show colonization has gone on for thousands of years, but somehow it is missing in our histories.
- Huge amounts of forestation; surface viewed, solar collecting, tunnels leading underground; Flying objects, and substantial amounts of water describe a planet ready for colonization, but there are no descriptions of early colonization potentials.

Let me say it one more time. People in UFOs are not from a distant galaxy, they are for earth. I know Gallaxians is an easy cop out for all the mysterious thing we found, but the are many problems with massive distance space travel while our earth is zooming through space as something like 50 thousand miles a second. All we need to recognize is when out brain began atrophying 5500 years ago, we forgot how to travel to our nearest neighbors and set up communities. We are just now making steps to regain our past glory and we will succeed. On Mars, it appears there are many things that can help us begin a new age of space exploration and give us a way to reduce crowding on earth without having to blow everyone up. When Neanderthal drew images of flying ships, he was not looking at aliens. They may have been his neighbors flying to Rome for dinner, or going to Venus to see the beautiful waterfalls in the massive mountain ranges. There are indications of life on the really cold planetoids of our solar system as well, but I'm talking about comfortable living on the moon, Venus and Mars.

About the Author

Steve Preston is a long lime author of scientific, esoteric facts. His books focus on the painful truths rather than whitewashed details that make us comfortable. If you are interested in the truth instead of comfort, please review other works by Mr. Preston as shown below.

His books include a wide assorttment of different subjects inlcuding Biblical History and proofs, the story of man's development, Ancient tecnology, New veiws of Physics and Biology, Anceint Wars, current fears and events. A partial list follows.

Development of Mankind

The First Creation of Man-book 1 History of mankind
The Second Creation of Man-book 2 History of mankind
The Creation of Adam and Eve-book 3 History of mankind
The Antediluvian War Years-book 4 History of mankind
Man After The Flood-book 5 History of mankind
A Closer Look at Ancient History-book 6 History of mankind
A New View of Modern History-book 7 History of mankind
The Twentieth Century and Beyond- Book 8 History of Mankind

Bible History, Correction, and Analysis

Abraham to Moses-First part of the Bible
Adam's First Wife-Story of Lilith
Adam to Abraham- Second Part of the Bible
Closer Look At Genesis- 200 ancient text confirm Genesis

Exploring Exodus- Reviewing the Details of "Exodus"
Errors in Understanding- Interpretations of the Bible
Expanded Genesis- Apocrypha and other Jewish texts
Exploring Genesis- Reviewing the details of "Genesis'
Incarnations of God- How often did God become Incarnated?
History Confirmed By The Bible- Science confirmation of the Bible
Moses Saved Egypt- How the Jews eliminated the Hyksos
Moses to Jesus- Third part of the Bible Series
Mysteries of the Exodus- Proofs of the Exodus
New look at the Bible- Questions in Interpretation
Old Testament Used By Jesus- Ancient Jewish texts
Understanding the New Testament-4th part of the Bible Series
Why the King James Bible Failed- Issues with KJB

Ancient Technology and Life

Anakim Gods- History of the Ancient Giant/gods
Ancient History of Flying- Ancient flying
Kingdoms Before the Flood- Pleistocene humans
Living on Venus- Venus before the Pleistocene Extinction
Martians- Ancient Life on Mars
Mysterious Pyramids- Who made the Pyramids?
Victory of the Earth- History of our Earth
Not from Space- UFOs are not from space.

Ancient and Modern War

America's Civil War Lie- Truth about the Civil War years
Behind the Tower of Babel- Story of the Bharata War
Driven Underground- Fear in the Bharata War
Four Armageddons- The 4 major wars that destroyed mankind
Six Deaths of Man- Destructions of mankind
World War Before- The Pleistocene War
World War with Heaven- The Angel and Anak War

World War Zero-The Bharata War
When Giants Ruled the Earth- History of the Titan Giants
Sex Crazed Angels- What caused the Heaven War?

Current Events and Fears

Allah' God of the Moon- Terror of Muslims
American School Disaster- fear in our country
Can We Save America? - Fear in the USA
Scythians Conquer Ireland- A History of Ireland
Fast History of MILES Training- Laser based Army training
Great American Quiz- Unusual details of American History
Make Your Own Global Warming
Truth About Phoenicia- The Evidence -First in America
Monsters are Alive- Post Pleistocene Monsters
Promote the General Welfare- Fear in USA
Our Very Odd Presidents- President review
Terror of Global Warming- Fake issue uncovered
The Antichrist- Many demonic possessed rulers
The Bad Side of Lincoln- Negative side of a great man
The Devil- Of Demons and their master
Vampires among Us- How Demons and Vampires are similar
Humans on Display- Slavery and Human Zoos

New Look at Physics

Amazing Technology- Pleistocene Technology
Anthropic Reality- We control our Reality
Consensus Science- Fake Science
Complex Earth- Truth behind Earth's development
Is Time Travel Possible? Science of Time Travel
Retiming the Earth- Eliminate of Nuclear Decay Errors
Releasing Your Consciousness- Beyond our SELF

Slip Through a Wall- How to walk through solids
Our 12-Dimensional Universe- New science of our Universe
Mystery of Photons and Light- Science of Photons
Of Heaven and Hell- scientific descriptions
Meaning of Life and Light- Detains of New Science
Vibrational Matter- New Science of Quantum Fluctuations

New Look at Biology

DNA of Our Ancestors- Tracing DNA of ancient man
God Didn't Make The Ape- New science on ape Evolution
Lizard People- Mutated People of the Bharata War
Creation and Death of Dinosaurs- Why Dinosaurs died
Races of Men- Tracing DNA of Humans
Tracing Cro-Magnon to Jesus- The third creation and mutation
Self, Soul, Spirit- Three components of Life
Self-Virtualization- New science of reality
True Happiness- Self Actualism and Beyond
Life Resonance- Unusual capabilities of men
Awaken the Departed- We can talk to the Dead
Biophotonics and Healing- How Photonics used in medicine

www.ingramcontent.com/pod-product-compliance
Lightning Source LLC
Chambersburg PA
CBHW050209230526
45470CB00001B/301